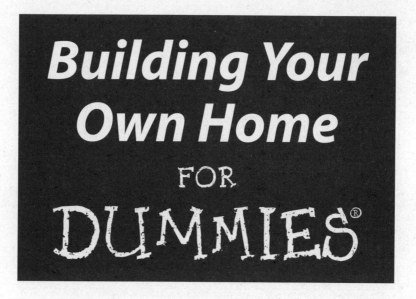

Building Your Own Home FOR DUMMIES®

by Kevin Daum, Janice Brewster, and Peter Economy

WILEY

Wiley Publishing, Inc.

Building Your Own Home For Dummies®

Published by
Wiley Publishing, Inc.
111 River St.
Hoboken, NJ 07030-5774
www.wiley.com

Copyright © 2005 by Wiley Publishing, Inc., Indianapolis, Indiana

Published by Wiley Publishing, Inc., Indianapolis, Indiana

Published simultaneously in Canada

For general information on our other products and services, please contact our Customer Care Department within the U.S. at 877-762-2974, outside the U.S. at 317-572-3993, or fax 317-572-4002.

For technical support, please visit www.wiley.com/techsupport.

Wiley also publishes its books in a variety of electronic formats. Some content that appears in print may not be available in electronic books.

Library of Congress Control Number: 2004117535

ISBN: 978-0-7645-5709-5

Manufactured in the United States of America

20 19 18 17 16 15 14 13 12 11 10

1O/QY/QS/QV/IN

WILEY

About the Authors

Kevin Daum, from Alameda, Calif., is founder and CEO of Stratford Financial Services, an INC 500 real estate finance company. Under Kevin's guidance, Stratford maintains a 100-percent approval rate on its custom home financing projects. Kevin has provided financing education for more than 20 years. He has underwritten loans for national institutions and managed real estate financing for entrepreneurs and celebrity clients, including Pat Sajak, Pat Boone, Dana Carvey, Phil Hartman, Jamie Farr, and Elvira, Mistress of the Dark. He has facilitated more than 850 custom home projects and is a recognized expert on the subjects of custom homes, real estate investment, and real estate management.

Kevin speaks regularly on the subject of real estate finance. Kevin penned and published the book *What the Banks Won't Tell You; How to Get the Most Out of Your Mortgage* (Grady Parsons) now in its second printing. He writes a featured monthly column for *Log Homes Illustrated* magazine and writes regularly for *American City Business Journals.* Kevin is actively involved with the Young Entrepreneur's Organization (YEO) and speaks on the subject of entrepreneurship and the arts. In addition to his entrepreneurial ventures, Kevin currently develops property and custom homes. He received his bachelor of arts degree from Humboldt State University.

To schedule Kevin to present to your organization, association, or conference, call Stratford Financial Services at 800-727-6050. For more information on the products and services provided by Stratford, visit its Web site at `www.stratfordfinancial.com` or contact Kevin directly at `kevin@stratfordfnancial.com`.

Janice Brewster is author of *Log Cabins* (Friedman/Fairfax) and *Cabin Styles* (Publications International, LTD). She currently edits *Timber Homes Illustrated* magazine and is former editor of *Log Home Living.* She has written articles for *The Washington Post, Cowboys & Indians, Catalina, Log Homes Illustrated, Timber Frame Homes,* and the National Association of Home Builders. Janice received her bachelor's degree from Mount Union College and her master's degree in magazine journalism from Syracuse University's S.I. Newhouse School of Public Communications.

Peter Economy, from La Jolla, Calif., is associate editor of *Leader to Leader,* the award-winning magazine of the Leader to Leader Institute, and author of numerous books, including *Managing For Dummies* (with Bob Nelson), *Home-Based Business For Dummies* (with Paul and Sarah Edwards) (both published by Wiley), and many others. He received his bachelor's degree (with majors in economics and human biology) from Stanford University and is currently pursuing his MBA at the Edinburgh Business School. Visit Peter at his Web site: `www.petereconomy.com`.

Dedication

This book is dedicated to all those information hungry consumers pursuing the American dream of home ownership.

Authors' Acknowledgments

Just like a custom home project, it took many people to make this book happen. We consulted experts in every area to make sure we included current, accurate information.

Kevin relied heavily upon the help and resources of the excellent staff at Stratford Financial Services, particularly Dawn Exline, vice president of client services. In addition Kevin wants to thank all the clients, builders, and architects that shared their experiences.

The authors also want to thank the following people for giving time, energy, and knowledge for the success of this book. Ahmad Mohazab and the expert team at Tecta Architects in San Francisco shared their knowledge and drawings for many chapters. Dan Ridings of Stonehenge Builders in Lafayette, Calif., spent hours of time providing technical information. Thanks go to The Original Lincoln Logs LTD., John Stetson, and Aaron Rosenbaum for their contributions of photos and Lorin George of Lorelco appraisals for her appraisal contribution. Thanks to Bud Davis of B. Davis construction for the special "Writing Place." Thanks to Scott Peloquin of BenefEx Benefit Consulting and Troy Collins of McKinley Financial for their experience in the financial area. Thank you to Charles Bevier, editor of *Building Systems Magazine,* for sharing his expertise.

Kevin, Peter, and Janice are also appreciative of all the people at Wiley Publishing, Inc. including Tracy Boggier, Joyce Pepple, Alissa Schwipps, Chad Sievers, Melisa Duffy, and Holly Gastineau-Grimes.

On a personal note, Kevin wants to acknowledge Mark Levy for his coaching as well as Tim Chrisman, Lisle Payne, and Dennis Erokan for their constant support and encouragement. He also wants to thank his friends and Forum at the Young Entrepreneur's Organization (YEO), which made this opportunity occur in the beginning. Finally Kevin acknowledges the love and support of his wife Deanna, son Spencer, and his parents Hal and Nancy Daum who showed him how to always put his clients' needs first.

Publisher's Acknowledgments

We're proud of this book; please send us your comments through our Dummies online registration form located at www.dummies.com/register/.

Some of the people who helped bring this book to market include the following:

Acquisitions, Editorial, and Media Development

Senior Project Editor: Alissa Schwipps

Acquisitions Editor: Tracy Boggier

Copy Editor: Chad R. Sievers

Technical Editors: Bob Gammache, (Carteret Mortgage www.nva-mortgage.com) and Dwayne Ganzel

Editorial Manager: Jennifer Ehrlich

Editorial Assistants: Nadine Bell, Hanna Scott

Cartoons: Rich Tennant, www.the5thwave.com

Composition

Project Coordinator: Adrienne Martinez

Layout and Graphics: Jonelle Burns, Carl Byers, Andrea Dahl, Joyce Haughey, Stephanie D. Jumper, Jacque Roth, Barry Offringa

Proofreaders: Laura Albert, Leeann Harney, Jessica Kramer, Carl William Pierce, Aptara

Indexer: Aptara

Publishing and Editorial for Consumer Dummies

  Diane Graves Steele, Vice President and Publisher, Consumer Dummies

  Joyce Pepple, Acquisitions Director, Consumer Dummies

  Kristin A. Cocks, Product Development Director, Consumer Dummies

  Michael Spring, Vice President and Publisher, Travel

  Brice Gosnell, Associate Publisher, Travel

  Kelly Regan, Editorial Director, Travel

Publishing for Technology Dummies

  Andy Cummings, Vice President and Publisher, Dummies Technology/General User

Composition Services

  Gerry Fahey, Vice President of Production Services

  Debbie Stailey, Director of Composition Services

Contents at a Glance

Table of Contents

Chapter 12: Excavation and Foundation: Getting a Solid Start 217

Chapter 13: Framing and Rough: So Much Goes Behind Those Walls! 231

Introduction

As you read this book, you most likely have the seed of a dream taking root in your mind. Your current home isn't all it could be. You've been daydreaming about a different place — one with more land, one by the water, or one with a gourmet kitchen. You've trolled the open houses in your area, but none of the homes really light your fire or seem to fit your lifestyle or your family. You want something that feels more like you.

The only way to get a perfect house "fit" is to design it specifically for you. No matter if your new home is a month away from completion or ten years down the road, you need this book.

In our work, we've seen plenty of people like you tackle the process of building a custom home. For some, the process is challenging, but enjoyable. For others, a custom home project becomes a nightmare that leaves them short on cash and long on anxiety. We understand the process and what it takes to move through it with as little stress as possible. In the pages that follow, we provide you with the very best advice our many years of experience have to offer.

No matter if your dream consists of a simple $150,000 house in the Midwest or a multi-million-dollar mansion in California, *Building Your Own Home For Dummies* is for you. This book can help you turn your dream of a custom home into reality without losing your shirt or your sanity. With this book and with some hard work and perseverance on your part, your dream of building, owning, and living in your very own custom home can become a reality.

About This Book

Thousands of parts and hundreds of tasks go into a custom home project. This book doesn't tell you how to install a toilet or hang a door (other *For Dummies* books cover those topics in detail), but it does tell you everything you need to know about creating a custom home from scratch. Where do you start? Who is responsible for what? How much will it all cost? These questions — and hundreds more — are what this book answers — and all in an easy-to-use reference that you can take with you anywhere.

We divide each chapter into sections, and each section contains information about some part of understanding the process of building a custom home, such as

- ✔ A comprehensive approach to financing your home project, before, during, and after construction
- ✔ The types of custom homes that are available — from log to timber frame to stick-built to modular
- ✔ A view from the loan officer's side of the desk
- ✔ A complete look at the inspection process during construction and what the inspector(s) will be looking for
- ✔ Thorough and helpful tidbits on how to successfully build your home and still have money left over

The great thing about this book is that *you* decide where to start and what to read. It's a reference you can jump into and out of at will. Just head to the table of contents or the index to find the information you want.

Conventions Used in This Book

We use the following conventions throughout the text to make everything consistent and easy to understand:

- ✔ All Web addresses appear in `monofont`.
- ✔ New terms appear in *italics* and are closely followed by an easy-to-understand definition.
- ✔ **Bold** text indicates keywords in bulleted lists or highlights the action parts of numbered steps.

We also include *spot-checks* in the Part III chapters to guide you in conversations with your contractor and help you make sure the construction process is going as planned.

What You're Not to Read

We've written this book so that you can easily find and understand information about building a custom home. Although you may be stuck on a deserted island and have plenty of time to read every word in this book, chances are you're not. So, we simplify it so you can identify "skippable" material. This information is the stuff that, although interesting and related to the topic at hand, isn't essential for you to know:

- ✔ **Text in sidebars:** The sidebars are the shaded boxes that appear here and there. They share fun facts, but nothing that's essential to the success of your project.

- ✔ **Anything with a Technical Stuff icon attached:** This information is interesting, but if you skip it, your custom house won't fall down.

- ✔ **The stuff on the copyright page:** No kidding. You can find nothing here of interest unless you're inexplicably enamored by legal language and Library of Congress numbers.

Foolish Assumptions

We wrote this book with some thoughts about you in mind. Here's what we assume about you, our reader:

- ✔ You've been sketching custom homes on napkins or doodling floor plans during business meetings. You've looked at your current home with a critical eye and have, at least once, sighed and muttered the phrase, "Someday. . . ."

- ✔ You're drawn to home-improvement stores, television shows, and books.

- ✔ You're desperately looking for a comprehensive guide that demystifies the home-building process by focusing on the information important for you the homeowner to know.

- ✔ You're willing to do some soul-searching in order to get your custom home right. You (and any significant others you may have) have decided that the only way to get the perfect home is to start from scratch.

- ✔ You don't live in a "money-is-no-object" world. You want to make educated financial decisions regarding the budget and long-term financing of your custom home.

- ✔ You want to be involved with the process but you'll rely on professionals to help you when you need it. Professional help may come in the form of a financial advisor or loan officer, an architect or designer, a plumber, or a landscaper. You're willing to assess your strengths and weaknesses and seek help when necessary.

- ✔ We assume that you'll hire a contractor in some capacity, as most people do. (We do provide some small tidbits of information if you want to be your own owner-builder, but the majority of this book focuses on building a custom home with a contractor.)

- ✔ You have the ability to keep an open mind and consider new approaches and information, even when they seem at odds with what you've always been told about the custom home and financing processes.

How This Book 1s Organized

This book is divided into five parts. Jump in wherever you want! The following sections explain what you'll find where.

Part 1: Getting Started: The 411 on Custom Home Building

Get up to speed on the basics of building a custom home. In this part, you figure out how to get your project organized and how to find property to build on. You start to envision your home and define its style. You get to know the role of the architect or designer and obtain an overview of the plan-approval process.

Part 11: All You Need 1s Dough: Financing Your Custom Home

Your project won't move from dream to reality without money. In this part, you can find the lowdown on using debt to your advantage and the construction loan process, including inside information on qualifying for the money you need to borrow. Read this part to understand why cash is king in getting your new home built.

Part 111: Hammers and Nails: The Construction Process

No, we don't expect you to build your own house with your own two hands. But wouldn't having some idea what those people are doing up on your roof or in your laundry room be nice? Find out the roles of the general contractor and the teams of subcontractors. Follow the construction process from excavation to framing to mechanical system installations to finish carpentry and beyond and use the provided spot-checks to make sure your contractor and subs are doing what they're supposed to do.

Part 1V: All the After Stuff

Just because the house is finished, you're not. Now it's time to plant and install your landscaping and, of course, move in. Also in this part, you see that money is still an issue — you need to consider how to manage the investment you've made in your new home. Your construction loan is closed, but it may not be too early to consider refinancing.

Part V: The Part of Tens

Like every *For Dummies* book, this part includes quick resources that provide plenty of information in an easy-to-digest fashion. Above all, this part shows that you aren't alone. Gain wisdom from other homeowners' trials and errors in the list of most common custom home mistakes and problems. Discover the best ways to lower construction costs. Use the list of best custom home resources to answer lingering questions or help you uncover wellsprings of useful information. We also provide an environmentally conscious list of ways to make your project green.

Icons Used in This Book

To make this book easier to read and simpler to use, we include some icons in the margins that can help you find and fathom key ideas and information.

These tidbits provide expert advice to help you save time and money in the home-building process.

This icon highlights important information to store in your brain for quick recall at a later time.

Avoid mistakes by following the sage words of advice that appear under this icon.

Although this information may be fascinating, it's not necessarily critical to your understanding the topic at hand. Feel free to skip it if you must.

Where to Go from Here

The process of building a custom home isn't linear. Not everyone starts with the purchase of a piece of land, for instance. Some people go to an architect first to help them create a floor plan. Others may jump right in with both feet, and be halfway through construction before they realize they need to borrow money in order to finish.

So, to reflect the nonlinear process of building a custom home, this book is decidedly nonlinear as well. We organize it so that you can dip in wherever you want and still find complete information. If you've already bought land and met with an architect, but don't know how you're going to pay for the project, for instance, go to Chapter 7 to read up on financing. Not clear who does what on the job site? Flip to Chapter 11 for information on general contractors and subcontractors.

If you're not sure where to go first, you may want to start with Part I. It gives you all the basic information you need to understand the process of building a custom home. From there, you can skip to sections that cover the subjects that seem most fuzzy to you now. Rest assured that when you've finished that section, you'll have a better grip on home-building reality.

Part I

Getting Started: The 411 on Custom Home Building

The 5th Wave   By Rich Tennant

"Look, why don't you and the Mrs. come to some final decision on where you want the house site, and then me and the fellows will come back and finish the driveway."

In this part . . .

Creating a custom home may be the biggest, most exciting project you have ever been involved in (yes, even more exciting than when you figured out static electricity in third-grade science class). As excited as you are though, you don't want to rush into it. In this part, we give you a general overview of what you're getting into. We also show you how to get organized and help you acquire land. Lastly, we help you decide on the type of home to build and walk you through the design and permit process with architects and designers.

Chapter 1

The Custom Home Process in a Nutshell

*M*ost people at some time in their lives desire owning a custom home. Some people are attracted to the thought of designing and creating something big from scratch. Others want to live in a new home that meets their specific needs instead of a house that looks like every other home on the block. Some people begin the custom home process by accident when they find a piece of land that inspires them.

More than 35 percent of new homes in the United States are custom homes. That means more than 300,000 custom homes are built every year. For each person building a custom home, five people are in the process of designing one. So you're in excellent company with many people dreaming about moving into a home designed and built just for them. Because custom homes are so popular, tons of resources are available to help you through the process.

But, like Rome, your new home won't be built in a day. The custom home process is lengthy, emotional, and expensive, without much consistency to it. Face it; custom homes require custom work and plenty of it! This work makes building a custom home challenging, and yet that extra work is what makes your project unique to you. You may feel overwhelmed at times, but by trusting in the experience of the professionals you engage in your project and keeping this invaluable book by your side, you can have a manageable project that delivers the custom home you have been dreaming of.

Where Do You Start? Preparing to Build Your Home

Believe it or not, the custom home process really has no standard starting place. There are some logical entry points such as finding land, but most often people start with a designed house they've had in mind for a long time. Where you start isn't important; what is important is for you to make sure that you have taken all the necessary steps to give yourself the best chance for success. The following list includes some questions you need to consider before committing time and money to this project. We discuss some of these issues extensively in other chapters (which we reference for you here).

- Where do I want to live?
- How long do I want to live in this house?
- How will I find land? (See Chapter 3.)
- How much money do I have to spend on this project? (See Chapters 7, 8, and 9.)
- How much extra time do I have to put into this project? (See Chapter 2.)
- How do I find the right resources to design my house? (See Chapter 4.)
- How do I find the right resources to build my house? (See Chapters 2 and 11.)
- Is my marriage/relationship strong enough to survive this process? (See your clergy or shrink.)

Don't make the assumption that any one person can give you all the information you need to prepare for this process. Contractors have one perspective on the process, and architects may have a completely different perspective. Do your homework and interview as many people as you can who are or who have been involved in the process. By talking to professionals and consumers and asking them to share their experiences, you can begin to get a clearer picture of the process ahead.

Kevin recommends to all his clients that they get organized before beginning the process. Sit down and assess how much time you can put aside each week to focus on the project. Consider making a specific day each week your day for working on custom home stuff. Also clear a space in your office or den to be "Custom Home Central." This way you always know where to find what you need for your project. (You can find other organizing tips for your project in Chapter 2.)

Money Makes the World Go Round — Paying for Your Home

We talk a lot about money in this book and with good reason. Custom homes require plenty of it. Your new home will probably be the most expensive item you have ever purchased. In fact, it may be the most expensive item you'll ever buy in your entire life. Custom homes cost more than production or tract homes because the materials aren't bought in quantities and the labor hired includes individual craftsmen. The results are worth it, however, and will last lifetimes.

Many people find it a challenge to get past the large checks they're writing. If you decide to use an architect, even the first check to the architect will probably exceed the biggest check you've ever written. The key to success with money in a custom home project is putting it in the right perspective. If your project costs $500,000, then what each piece costs isn't important as long as it equals $500,000 or less.

When you buy a new car, you don't argue over how much you spent for the alternator or the exhaust system. You look for the car to meet the price of your overall budget. Use the same logic when buying your custom home. Look for the best price on each item, but look at it in perspective to the entire budget. You'll do better on some items and worse on others, but if it fits your finances, then you're in good shape.

Asking yourself about affordability

Of course you have heard horror stories about custom home projects that have gone seriously over budget. The projects go over budget for many reasons, but usually the main culprit is that the potential homeowners didn't spend enough time determining what they could afford. Obviously, if you're building well below your means, then going over budget is easily rectified by using your own cash. But running out of money is the No. 1 cause of custom home disasters. Before you start the custom home process, you seriously need to consider the following:

- ✔ **What can you physically pay?** Take stock of your cash on hand, equity in real estate, and available cash from other resources. Make a firm decision how much money you're willing to put toward the project. Chapter 7 can be a big help. You also need to get a rough idea of how much borrowing power you have to help establish a limit for your budget when added to your available cash. We provide tools and Kevin's expert financing assessments in Chapters 8 and 9.

✔ **What can you emotionally pay?** Just because you have the money and the borrowing power doesn't mean you really want to spend it all. Think carefully and discuss with your spouse what your limits are for making payments and how much *liquidity* (or cash) you need in the bank to help you sleep at night when all is said and done. Make sure you take into account tax deductions and interest earned on investments when analyzing your monthly cash flow. After you have found that emotional limit, you can design your project to fit your comfort zone.

✔ **What is your cushion and tolerance for risk?** Like we say again and again throughout this book, building a custom home is a complex process. You need to consider many variables beyond your control, and then realize that the project can go over budget even if you do everything right. You can certainly get good solid estimates, but ultimately you won't know what this home will cost until it's finished and you total up the receipts. Make sure you have addressed the "what if?" issues thoroughly. Talk about how you'll cover things financially if the market turns sour — devaluing your property — or the cost of materials rise. Decide what safety money (such as your 401(k) or retirement fund) you're willing or unwilling to tap into.

The more you talk about financial issues related to your custom home project, the more likely you are to resolve problems before they happen. Optimism in a custom home project can get you into trouble every time. The best approach is to examine every possible risk and make contingency plans for every potential problem.

Them that has the gold makes the rules: If you finance, the bank will dictate process

Most people don't have all the money for a custom home sitting in their bank account. Even if they did, putting it all into the project wouldn't be a good idea, as we explain in Chapter 7. Like it or not, you'll probably have a financial partner in this project in the form of a construction lender or bank. The good news is construction lenders have the same objectives you do.

✔ They want to make loans for custom home projects. (That's how they make money.)

✔ They want the house to be completed on time.

✔ They want the house to be completed on budget.

✔ They want the house to be completed in a workmanlike manner.

Furthermore, the following tidbits can save you some arguments and frustrations when working with construction lenders:

- ✔ They don't believe a house is worth exactly what it costs.
- ✔ A larger loan makes you a riskier borrower, not a better customer.
- ✔ You aren't entitled to any loan.
- ✔ They aren't required by law to loan you any money.
- ✔ They dictate how the money is handled throughout the process.

Accept the fact that if you want to use a lender's money, you have to play by its rules. Most of these rules weren't made arbitrarily. They're designed to protect the financial viability of the project and protect the lender in the unlikely event of a *foreclosure,* which is the act of taking back the home in case you default on the payments or the construction contract. The guidelines and procedures are based upon statistical and anecdotal problems and failures that occurred with the lender in the past. Unfortunately, sometimes you pay for the sins of those before you.

Put yourself in the lender's shoes. If you were loaning a friend 80 percent of the money to build his home, you would want a few protections in place and a little control over the money as well, right? If you get to know how lenders see the project, which we explain in detail in Chapters 8 and 9, you can easily navigate the approval process as well as the funding process (see Chapter 10). This approach can make for a smoother, happier custom home project.

Introducing the Custom Home Life Cycle

The first step to beginning the process is looking at all the pieces and how they go together. Your new home has a number of individual projects and transactions necessary to complete it. Your new home also needs an army of people with their expert work and services. This section breaks down in an approximate order each person required to get through the process. Then we outline each step necessary to go from land to landscaping.

It takes (more than) two to tango — A quick guide to the players

The following list is a guide to all the individual players involved in the custom home process. You may or may not use them all; their roles can vary depending on your region and your project's scope. The order of need may also change depending upon where you start in your process.

- ✔ **Financial planner and/or certified public accountant (CPA):** If possible, start the custom home process by carefully assessing your finances; a financial planner or CPA can help make sure you can afford this project.

- ✔ **Real estate agent:** You may need a real estate agent to help you find and purchase a lot, as we discuss in Chapter 3. She also plays a role when it's time to sell your existing home.

- ✔ **Loan officer:** Your loan officer needs to be involved throughout the entire process. You may need to start with a refinance or credit line to get liquid, as we discuss in Chapter 7. You want to finance the land (see Chapter 3) and do it consistent with the construction financing (see Chapters 8 and 9). Finally, you still may need another refinance after the project (see Chapter 16). Your loan office can help you through these steps. Lucky for you, Chapter 8 has good advice on picking the right loan officer.

- ✔ **Developer or landowner:** The land you buy has to come from somewhere. If you're buying in a subdivision from a developer, you may deal with a sales office. Or you may end up buying from a landowner that has had the property for generations.

- ✔ **Escrow officer or attorney:** Your state determines who administers the closing of your escrow, but either way, this person makes sure the title papers and insurance are all ready for you to take ownership.

- ✔ **Architect and/or designer:** Architects and designers design and draft plans for the house. Architects are licensed; they'll coordinate technical specifications for the house that may be beyond the scope of a designer. The architect can also guide you through the permitting process. (Chapter 5 can help you decide if you need an architect, and Chapter 6 provides the ins and outs of the permitting process.)

- ✔ **Log or timber frame dealer:** If you're building a kit home (see Chapter 4), you'll work with your dealer for the design process as well as the purchase of your materials package.

- ✔ **Contractor/builder:** You need to decide whether you need this person or if you'll rely on yourself to drive the construction of your new home (see the "Being an Owner-Builder: More Power to You!" section, later in this chapter, if you may want to be your own owner-builder). We give you tools for working with your contractor in Chapter 11.

- ✔ **Surveyor:** This person makes sure you know where your land begins and ends — a necessity for designing the house.

- ✔ **Soils engineer:** In many states, such as California, your foundation depends upon the report issued by this person.

- ✔ **Well/septic engineer:** If you're building in a rural area, you need this person to design and certify your water and sewage systems.

- ✓ **Planning department:** Your house needs to meet your neighborhood's zoning requirements before you get permits. This department enforces the zoning. (See Chapter 6 for details.)

- ✓ **Design review committee:** You can't always build what you want. This committee dictates what it wants to see in your design. (Look in Chapter 6 for more information.)

- ✓ **Building department:** Everything must meet code, and this department checks your plans before issuing permits. (See Chapter 6.)

- ✓ **Appraiser:** The lender won't approve a construction loan without an appraisal estimating the finished value. (Check out Chapter 9 for more information.)

- ✓ **Insurance agent:** Chapter 2 spells out all the insurance you need for the project. This person provides the goods — he'll be busy.

- ✓ **Material suppliers:** Sticks and stones all have to come from somewhere. Some projects have many sources. (See Chapter 11.)

- ✓ **Subcontractors:** Each one is an expert . . . just ask them. Artisans and craftspeople build each different system in your house. Chapter 11 tells you how to work with them. Chapters 12, 13, and 14 explain what they do.

- ✓ **Laborers:** Somebody has to do the grunt work on the job. These guys and gals work the hardest and get paid the least.

- ✓ **Building inspectors:** The building department checks up at various stages of construction to see that you're building in line with regulations. (Look in Chapter 11 for more details.)

- ✓ **Disbursement agents:** The lender assigns someone to make sure you get money when you need it or to solve problems with getting money from the lender. (You can find more on these agents in Chapter 10.)

- ✓ **Bank inspectors:** The bank won't give you money unless work has been done. These people come out to the property monthly or at various stages to make sure the work is complete. (Chapter 11 has more.)

- ✓ **Landscaper:** Usually the last part to go in but sometimes the landscaper designs the landscaping at the beginning. This person makes the yard green with your green. (Check out Chapter 17 for more info.)

- ✓ **Mover:** After all this work and trouble, the last thing you want to do is make 20 trips with the minivan. Let the movers do the work for you. (Turn to Chapter 15 for specifics.)

- ✓ **Decorator:** If you have any money left at the end, you'll have plenty of furnishings to spend it on. An interior decorator can help.

Although your architect or contractor may manage some of these relationships, ultimately you'll need to coordinate all these people in order to complete the project. You're going to meet many new people in this process, so put on your best smile and get ready to shake a lot of hands.

So many tasks, so little time —
50 steps to a custom home

You're probably wondering why the custom home process has so many people involved. The simple answer: A custom home process has tons of tasks that need to be done. Although each home-building process may have some variation in the stages based upon factors such as location and weather, for the most part, the process moves in a step-by-step fashion.

The following list shows how a typical custom home process moves forward. The chapter references direct you to detailed discussions later in the book.

1. Decide you're ready to tackle the custom home process. (See Chapter 2.)

2. Meet with financial experts and get organized. (See Chapter 2.)

3. Prepare cash flow with financing on your existing house. (See Chapter 2.)

4. Find land and make an offer. (See Chapter 3.)

5. Obtain land financing. (See Chapter 3.)

6. Close escrow on the lot. (See Chapter 3.)

7. Get surveys and soil reports. (See Chapter 3.)

8. Get well and septic approvals if required. (See Chapter 6.)

9. Interview and pick an architect, if applicable. (See Chapter 5.)

10. Create the house's preliminary design. (See Chapter 5.)

11. Get zoning and design review approval. (See Chapter 5.)

12. Pick all your fixtures and materials. (See Chapter 5.)

13. Submit the plans for building approval. (See Chapter 5.)

14. Make the required plan changes. (See Chapter 5.)

15. Put the plans out to bid with contractors. (See Chapter 2.)

16. Interview and choose a contractor. (See Chapter 2.)

17. Apply for a construction loan. (See Chapters 8 and 9.)

18. Get an appraisal based on future value. (See Chapter 9.)

19. Get final approval for permits and pay fees. (See Chapter 6.)

20. Close escrow on the construction loan. (See Chapter 8.)

21. Set up disbursement account. (See Chapter 10.)

22. Set up communications with contractor and subs. (See Chapter 11.)

23. Prepare the building site for work. (See Chapter 12.)

24. Grade and/or excavate the property. (See Chapter 12.)

25. Trench for foundation, water, and sewer. (See Chapter 12.)

26. Pour the concrete for foundation and let cure. (See Chapter 12.)

27. Frame the exterior. (See Chapter 13.)

28. Frame the interior. (See Chapter 13.)

29. Install the windows. (See Chapter 13.)

30. Install the fireplaces. (See Chapter 13.)

31. Install the rough HVAC. (See Chapter 13.)

32. Install the rough plumbing. (See Chapter 13.)

33. Install the rough electrical. (See Chapter 13.)

34. Install the roof. (See Chapter 13.)

35. Install the outer sheathing. (See Chapter 13.)

36. Apply the exterior siding or stucco and paint. (See Chapter 14.)

37. Install the drywall. (See Chapter 14.)

38. Install the cabinetry and millwork. (See Chapter 14.)

39. Install tile, counters, moldings, and finish carpentry. (See Chapter 14.)

40. Install the doors. (See Chapter 14.)

41. Paint the interior and finish woodwork. (See Chapter 14.)

42. Install the plumbing fixtures. (See Chapter 14.)

43. Install the electrical fixtures and hardware. (See Chapter 14.)

44. Install the flooring. (See Chapter 14.)

45. Request the final loan disbursement. (See Chapter 15.)

46. Request final inspection and receive certificate of occupancy. (See Chapter 15.)

47. Roll to permanent financing. (See Chapter 16.)

48. Install the landscaping, including deck, pool, spa, and so on. (See Chapter 17.)

49. Sell your old house. (See Chapter 15.)

50. Move in. (See Chapter 15.)

Figure 1-1 shows photos taken through a number of stages to give you an idea of what a home in progress looks like.

Figure 1-1: A house in progressive stages of construction. From site prep through foundation, framing, and exterior work, this home took more than eight months to build.

Courtesy of Aaron Rosenbaum

Patience is a virtue — A true timeline for building your home

Asking how long it takes to build a custom home from start to finish is a bit like asking the question "How long is a piece of string?" The obvious answer of course is "It depends." So many factors can affect the time frame that the overall project can stretch from six months to six years. Kevin often has clients who come to his office asking if they can move in by Christmas, to which he always responds, "Absolutely, as long as you don't care which year!"

Over the years we've seen patterns for the time it takes to complete each phase. The main point is to be flexible. You want to have a house you love for the rest of your life rather than years of regret because you rushed everything. Here are some typical rough timelines for the process based on Kevin's 20 years' of experience.

> ✔ **Land acquisition:** This step depends upon the availability of land in the area you desire. Land is hard to find, so pinpointing the exact time is difficult. Most of Kevin's clients look for land for three to nine months before finding something they like. Purchasing the land, including the escrow and due diligence periods, can take anywhere from 30 days to six months.

✔ **Home design and approval process:** This stage mostly depends on how picky you are and your financing considerations. Local government efficiencies can play a factor as well. Figure at least three months. The design and approval process requires that everything goes perfectly and you can quickly make your choices. Kevin has projects that have taken more than two years to get through this phase.

✔ **Construction:** This stage covers the project's scope and the availability of labor. You can use the construction lenders as a guide. Most lenders provide 12-month construction loans. Smaller houses or *kit homes* (homes where all materials are supplied as a kit, such as log homes) may go up in six to nine months. Large detailed mansions may need 18 months.

✔ **Landscaping and move in:** This one is all up to you. After the house is complete, you can relax, although you may be required to finish landscaping in some neighborhoods within a year of completion. Most finish within six months.

Being an Owner-Builder: More Power to You!

When you talk about building a custom home, people often assume you're planning on pounding hammers and nails yourself. Hardly anyone does the actual construction on their own custom home project. Many people, however, do consider acting as their own general contractor. Still, doing so is such a large undertaking that only about 20 percent of all custom homes are managed by *owner-builders*. In many of these cases, the owner is a contractor or already has some amount of construction experience. This factor isn't necessary, but it can make a big difference in the ultimate success of the project.

Even though the primary motivation for considering becoming an owner-builder may be saving money, the real issues to consider are time and management experience. This project will be one of the largest undertakings of your life, even with a contractor. Consider the following questions in exploring the owner-builder subject:

✔ How is my security at my current job?

✔ Do I have extra time and a flexible schedule?

✔ Can I make more money at my job with the time I spend on the home?

✔ Do I have a good understanding of the construction process?

✔ Do I have extra time to train myself on the process?

✔ Am I good at managing people and projects?

✔ Do I have a good eye for quality of construction?

✔ Do I have access to good resources?

✔ Am I good at problem solving?

✔ Am I good with multitasking and constant change?

✔ Am I well organized?

✔ Am I good at managing finances and budgets?

✔ Will my spouse and kids stay with me if I mess up the project?

If you honestly answered no to any of these questions, then you probably need to hire a contractor (see Chapters 2 and 11). Most owner-builders are gambling that they can do a job equal to or better than an experienced, licensed contractor, thereby saving the cost of that contractor. Although an owner-builder may end up saving money, you need to weigh the risk of that gamble against the money you might save. If you're wrong, it could cost you far more money than you planned to save in the first place.

One option if your answers were somewhat mixed is to hire an owner-builder consultant. One company called Ubuildit (www.ubuildit.com) offers expert consulting and procedures to guide you through the construction manage-ment process. The company charges you consulting fees and offers you prod-ucts and services that are marked up, but the costs can be significantly less than a contractor's fees. Ubuildit is a good alternative for saving money and shortening the learning curve; however, you still need to have the time and the management skills to make for a successful project.

Analyzing the truth about savings

The biggest motivation for being an owner-builder is the supposed savings. Ordinarily, a contractor makes money from charging a percentage on top of the cost of labor and materials used in the project; this fee or markup can be anywhere from 12 to 35 percent depending upon what and where you're building. Generally, more established contractors work on higher margins where younger contractors with less experience may work for less.

Where materials are concerned, the discount suppliers such as Home Depot have made construction supplies available to the consumer at contractor prices, which can be real savings if you're satisfied with the selection avail-able at these stores. If you're building with more elaborate materials and fix-tures, the contractor may have access to wholesale pricing that allows him to make some money without you having to pay more. In some cases he may be working on a lower margin and may be able to save you some money on items with a high retail markup.

With labor, you'll be subject to the prices and availability of the subcontractors in the marketplace. If the market is busy, pricing will reflect a direct supply-and-demand relationship, pushing prices up. If you have no preexisting relationships with any subs, you'll end up paying the full price for their time. If they're unable to work into your schedule, you may have other costs that come from delays on your project while waiting for the subs to become available.

Finding and managing subs

Hiring and managing subs is the hardest part of being an owner-builder. Meanwhile, a contractor has the advantage of having long-term regular relationships with subs. An experienced contractor has spent years finding framers, plumbers, carpenters, and others whom he trusts to be timely, efficient, and good craftsmen. If they've worked together for a long time, they know how to work together, and the contractor will know when to ask for favors.

Hiring each sub is a new experience in negotiation, management, and quality control. Overcommunicate with everyone on the job to keep it running smoothly. Keep your eyes open. You probably won't know if you picked the right sub until she is finished and she has been paid. (Check out Chapter 11 for more information about working with subs.)

Financing implications

One other challenge with being an owner-builder is the financing. Most conventional construction lenders frown on owner-builder projects. They have three basic reasons for being concerned:

- ✔ The bank is afraid the project might not be managed effectively causing it to exceed the allotted time frame and budget.
- ✔ The bank wants to be sure your job and income won't be negatively impacted by the time demands of this project.
- ✔ In case of foreclosure, the bank doesn't want to have to find and hire a contractor to finish the home.

For these reasons, many banks who lend to owner-builders do so with stricter requirements than for regular construction loans, such as loaning less money relative to appraised value or requiring full income documentation. Others allow owner-builder financing only if you're a general contractor, or at the very least they require someone with construction experience as a site supervisor. Private sources for owner-builder construction loans are available, but they can be expensive and don't generally have permanent loans attached like the single-close loans we recommend in Chapter 8.

Chapter 2

Preparing for the Process

. . . .

. . . .

Any time you undertake a multistepped project, you have a greater risk of something going out of control. The good news is that you can prepare yourself for the chaos and craziness that is bound to happen in your construction project.

In this chapter, we help you set up some simple systems for managing the people and tasks involved in the custom home process. We walk you through a short analysis of your finances so you can create a budget. We take you through the decision process of selecting a contractor and also help you to understand your insurance needs for the project. Finally, we provide you with several tips on how to keep the project a happy experience.

Organizing and Documenting

The custom home process is chock-full of enough paperwork and procedures guaranteed to give bureaucrats chills. Now is the time to be honest with yourself. Are you truly an organized person? If so, this section is simply a series of reminders and ideas for you to embrace. If not, don't get intimidated by the challenges ahead of you. Find someone in your family who is organized or even hire someone to help you prepare for the large organizing task ahead. One good resource is the National Association of Professional Organizers at www.napo.net. You can also check out Wiley's *Organizing For Dummies* by Eileen Roth and Elizabeth Miles.

Building a workbook and portable file system

To start with a difficult project like you're thinking of undertaking, you need a central place to store all the original paperwork you're about to accumulate. Because each transaction creates its own set of paperwork, you want to get organized or else you'll end up drowning in all that paper. A typical construction project usually generates enough paperwork to fill a two-drawer file cabinet. You not only need to store all this paperwork, but you'll also need to easily retrieve it throughout the process. Use the following suggestions for setting up categories for your filing system:

- ✔ Architecture and design
- ✔ Contractor communication
- ✔ Contracts
- ✔ Financing
- ✔ Invoices
- ✔ Land purchase
- ✔ Materials information
- ✔ Paid receipts
- ✔ Permits and approvals
- ✔ Subcontractor communication
- ✔ Warrantees

Many people start out with a single notebook and find out it fills up very quickly. We recommend using a permanent and portable system instead. Utilize the following efficient, step-by-step method for having pertinent information at your fingertips whether you're at home, in your office, or at your construction site:

1. **Create a loose-leaf binder with dividers for the categories in the previous bulleted list.**

2. **Decide after looking at each document whether you may need it at the site. If so, make a copy.**

3. **File one copy in your file system at home.**

4. **Place the other copy in your binder in the corresponding category to the home file system.**

5. **Review your binder each day, adding the necessary documents from your file system.**

More and more people in the construction industry are using e-mail for communication. People collaborating on your project can easily pass along and share invoices, designs, pictures, and memos via e-mail. The great thing about e-mails is that they're easy to store without taking up any space in your file cabinet. If you use Microsoft Outlook or Lotus Notes, you can set up different folders for all the subjects and people you communicate with. This way you can easily reference prior communication and share it if needed. If you don't currently use e-mail, take the time now to figure it out. Doing so can make your custom home project run more smoothly. (Check out *Outlook 2003 For Dummies* by Bill Dyszel, published by Wiley, for help setting up and figuring out how to e-mail.)

Calendar and communication — Your PDA is your friend

Keeping your project on schedule is a major project in itself. You and your contractor need to coordinate all the actions in a construction project. For example, your electrical systems can't be installed until the framing is complete, and the house can't be framed until the foundation is installed. With so many people dependent upon the time frame of others, you need a simple way to keep track of everything, even if your contractor is managing the schedule.

Proactive communication is probably the single most important factor for a successful custom home. Make yourself easily available with a mobile phone so your contractor or architect can reach you when she needs you. Be prepared to respond to messages promptly (otherwise, if your crew runs into a snag, the project could sit in limbo while your contractor waits for you to check in and return messages — costing you time and money). If you're a recluse or shy when it comes to dealing with people, you may need to adjust your lifestyle and contact-management approach until your project is completed.

Staying in close contact requires you have immediate portable access to phone numbers for your contractor, architect, loan officer, and other key players. For those of you electronically minded, you can utilize great techno tools to help you. A personal digital assistant (PDA) is a small handheld computer that can store all your phone numbers as well as your calendar. PDAs cost from $100 to $500 and can transfer information back and forth from your computer, which helps if you're communicating by e-mail or managing your calendar electronically. Both Palm (www.palm.com) and Microsoft (www.microsoft.com) make software that is compatible with most computers. You can take time to figure out how to use these devices by reading Wiley's *Palm For Dummies,* 2nd edition, by Bill Dyszel, or *Pocket PC For Dummies,* 2nd edition, by Brian Underdahl.

A number of phones are also available today that combine the functions of PDAs with a fully functional mobile telephone. They're pricier and bulkier than regular mobile phones, but they allow you to combine the functions of two units. Check with your local cellular phone provider for details.

If electronics aren't your thing, we highly recommend a Franklin Planner from www.franklincovey.com. They come in many sizes that can also serve as your planning notebook. You can keep all your contact information, as well as your calendar, with a pencil and have it ready whenever you need it. And if you use a planner, your batteries will never go dead because — unlike PDAs and other portable electronic devices — planners don't use batteries!

Being the bean counter — Keeping track of your finances

Even though you may have a contractor and bank involved, ultimately, the job of managing the finances falls on you. You need to keep track of every dollar spent as you go, or you could have a very unpleasant surprise — running out of money in the middle of the project.

We recommend setting up a separate bank account early on for everything construction related. Setting up this account can help remove confusion and allow for easier record keeping. Keeping a file for each vendor and tacking down invoices and receipts in chronological order also makes life easier when looking for something later. Loose papers become a nightmare when you need something quickly. (See the "Building a workbook and portable file system" section earlier in this chapter for specific tips on keeping the files organized.)

Managing the finances is an easier task (just like keeping your schedule organized) if you're computer literate. You could manage your finances with a simple bookkeeping ledger book, but you can keep track of money in and money out in all the different categories of the build smoothly and efficiently with the help of an accounting software program such as QuickBooks by Intuit (www.quickbooks.com). QuickBooks offers a special construction version designed for contractors that can serve your needs well. The version is a little pricey at a few hundred dollars, but that's a small price to pay for effective money management. Wiley's *QuickBooks For Dummies* by Stephen L. Nelson can guide you through the software quite nicely.

Shopping and sharing — Collecting material information

The biggest assignment you have in this project is . . . to go shopping! You need to choose hundreds of items for this house, and the sooner you start saving pictures and catalogs the better. (We provide a basic list of choices to be made in Chapter 5.) Magazines, catalogs, and the Internet are your best bet for finding hardware and fixtures.

Electronic storage of pictures makes for easier communication with your architect and contractor. A scanner can be a useful tool for cataloguing pictures from magazines to store on your computer. Then you can e-mail the scanned images to your architect and contractor or burn the images to a CD and share the disc with them.

Budgeting Your Project

A budget for a custom home project is a living breathing animal. It will grow and shrink many times before the house is finished. Okay, it usually grows more than it shrinks, but the point is that it changes — a lot! You do have to start somewhere. This section can help you create a preliminary budget to get started. (We also provide more specific budgeting help in Chapter 9 where we explain how a lender budgets your project. If you're financing this project through a lender, then the lender's budget takes precedence over yours, so you need to get the two in line as soon as possible.)

Budgeting at the beginning of the custom home process is a bit of a chicken-and-egg process. You have to balance out the cost of the house with what you can afford. The problem is that the house may require more money than you have, which you can't figure out until you design the house and so on. The best method is to evaluate the two issues separately and then work to a compromise.

Looking at your finances and cash flow

Chances are your lender will heavily influence your budget by determining an amount to lend you. But just because the lender says you can afford a million dollars doesn't mean you're comfortable spending that much money or making those payments. The best way to create a budget you can live with is to work with your certified public accountant (CPA), financial advisor, and loan officer to assess how all factors will impact your finances. Make sure you discuss and take into account the following:

- ✔ Cash on hand
- ✔ Capital gains issues
- ✔ Current tax bracket
- ✔ Diversification of assets
- ✔ Intended length of time owning the home
- ✔ Long-term investment strategy
- ✔ Property appreciation
- ✔ Tax deductions for interest and points

Armed with this information, you need to arrive at a comfortable payment that a loan officer or loan calculator can translate into a loan amount. You can find a variety of these calculators at www.mortgage-calc.com.

The mortgage information set out by these calculation sites is basic information and doesn't totally apply to construction loan qualification. These sites can give you a rough estimate to work within preliminary stages, but you need to speak to a loan officer that is a qualified construction loan specialist to be assured you meet construction loan qualification. (You can find specific information for construction loan underwriting and for finding a loan officer in Chapters 8 and 9.)

After you have a loan amount, you need to account for the cash available. As we discuss in Chapter 7, you need a good amount of cash to run this project. You may not want to spend it all, but cash is your surest way of keeping a custom home project running smoothly. Don't forget to include money you can take out of your existing house through a credit line or refinance. After you have decided how much of your cash you want to spend without being reimbursed by the construction loan on your project, add it to your loan amount estimate for the total estimate of your custom home budget.

Just because you can afford a large budget doesn't mean that the property will support the amount you want to spend. Many other factors can impact your budget later in the process, such as requirements of your property and sales in your neighborhood. (We discuss these elements extensively in Chapters 3 and 5.)

Defining "dollars per square foot"

Many different people use the "dollars per square foot" term many different times during your home-building process. Interestingly enough, however, no one uses a specific widely accepted definition. A real estate agent may state dollars per square foot as the sales price of the home divided by the square footage, including the land. A contractor may or may not include items such as permits or financing in his estimates of dollars per square foot. After you have bids, the term actually becomes meaningless, but during the early stages of your project, you need a common understanding of what it means.

To decipher your dollars per square foot quotes, you have to define dollars and square foot the same way for each person you work with. Then you can make sure everyone is on the same page in every relevant conversation. The following sections outline the approach Kevin takes with his clients.

Step 1: Define square footage

First, create a definition for "square foot." *Square footage* for this purpose needs to include all living space enclosed by walls that is completely finished. Your definition of square footage needs to include the square footage for the following:

- ✔ Bathrooms
- ✔ Bedrooms/closets
- ✔ Den
- ✔ Dining room
- ✔ Family and great room
- ✔ Fully finished basement
- ✔ Guesthouse
- ✔ Hallways and entryways
- ✔ Home theater and/or game room
- ✔ Kitchen/laundry room/pantry

Add the square footage together, and this total serves as your definition of the total square footage. However, your definition of square footage doesn't include square footage for the following:

- ✔ Attached decks
- ✔ Garage
- ✔ Patios
- ✔ Unfinished basement
- ✔ Workshop buildings

Step 2: Define dollars

Now that you have a total square footage number, you need to define the dollars necessary in the budget to define dollars per square foot. You don't want to include all the construction costs in this dollar amount; many costs need to be evaluated independently. The following are the costs that Kevin excludes from this part of the calculation:

- Financing
- *Hardscaping* (unattached decks, pools, fences, and so on)
- Land
- Landscaping
- *Soft costs* (permits, plans, and fees)

So what's included in your definition of the dollars? Mostly labor and materials construction costs for all the living space we mention in the "Step 1: Define square footage" section are included plus a few other construction costs such as the cost of

- Attached decks and patios
- Driveways
- The garage
- Unfinished basement space
- Walkways

Take all your cost estimates and add them together to create a total dollars number.

Step 3: Calculating dollars per square foot

As soon as you establish the total dollars number, divide the total dollars from Step 2 by the square footage number from Step 1 to establish your dollars per square foot. Easy, huh?

Alternatively, you can take your total dollars construction budget minus the excluded items from Step 2 and divide by the square footage number to determine how much per square foot you have available to spend.

For example, if you have a budget of $350,000 for total dollars available and your square footage is 2,500 square feet, your budget would be $140 per square foot. You could then tell a contractor that you can only spend $140 per square foot for construction of the house not including land, soft costs, financing, hardscaping, and landscaping, but that price must include the garage, driveways, walkways, attached decks, and any unfinished space.

Using a budgeting template

To put all this information together, Kevin's company, Stratford Financial, suggests a little template to make everything easier. You need to do your own research to fill in the spaces for your project, but the following is a sample preliminary budget for a typical custom home project that you can scratch out on any napkin.

Funds available

Add your available cash and the loan amount for your total budget.

Cash	$175,000
Loan amount	<u>$650,000</u>
Total budget	**$825,000**

Cost-to-build

Add all your costs together for your total cost.

Land	$200,000
Soft costs (permits plans and fees)	$40,000
Hard costs ($160 per 2,650 Sq. Ft.)	$424,000
Financing	$35,000
Landscaping	$40,000
Hardscaping	<u>$25,000</u>
Total Cost	**$764,000**

This template can give you a starting point for budgeting, but you do need to educate yourself on each of these line items to get a real picture of your project's costs. (Check out the table of contents to see where we discuss each topic for more information.)

Hiring a Contractor

Most people building a custom home end up hiring a general contractor to do the job. In fact, 80 percent of custom home projects have a general contractor involved in some capacity. Many people find great comfort in having someone with experience manage the job while they earn the money to pay for the project. If you're considering being an owner-builder and not using a contractor, check out Chapter 1 to see if you're truly up to the task.

Every contractor has a standard contract and practices she uses with her business. Remember that you have the right to negotiate and compromise on issues before entering into any contract. Of course, the contractor also has the right to decline the job.

There is no set time to engage your contractor as long as it's a minimum of 60 days before you start construction so she has time to get everything ready for the build. Many people opt to engage a contractor much earlier in the process so that the contractor is actively involved during the design process.

After you and your contractor establish a contract and you're on your way, you and the contractor need to work as a team to build your custom home. In this section, we look at the contractor selection process. (You can find more information on managing your contractor relationship in Chapter 11.)

Understanding the contractor's role

In most cases, the contractor doesn't handle the hammer-and-nail part of your custom home project. Although some contractors may participate in parts of the actual construction, his primary job is to manage the workflow and project materials and make sure everything is happening in a timely and workmanship-type manner. The following list includes your contractor's major responsibilities:

- Obtain the final permits
- Manage the production schedule
- Source and buy the materials
- Hire and manage the subs
- Keep the site safe and clean
- Manage the inspections
- Manage the budget

> ✔ Keep the consumer informed
>
> ✔ Effect quality control
>
> ✔ Problem solve

Not every contractor will handle all these items. When you hire your contractor, discuss with her how much you'll handle yourself directly and what she'll handle. Good communication is the key to making sure that you don't waste any time or effort doubling up on tasks.

Getting bids — Comparing apples to apples

If you didn't start the custom home process with a contractor in mind, you'll probably give your plans and specifications or *specs* to a few different contractors to get an estimate. This process is called putting your plans *out to bid*.

Finding contractors to bid on your project is as easy as asking friends and your architect for referrals. If they aren't providing names, you can drive through the neighborhood and look for construction signs on houses being built. The Internet is also a good resource. Simply type the name of your city and the word "contractor" into a search engine such as Yahoo! or Google. Dozens of referral sites pop up for you. If the building market is busy, you may have to work a little harder to find contractors who can bid in your time frame. You want at least three contractors to bid on your project if possible. The trick is making sure you can compare the bids to each other.

The best way to compare the bids is by making sure your plans are as complete as possible. In Chapter 5, we discuss doing all your design work and material selection before the bid process. This extra work can delay your project early on but is crucial for comparing truly accurate bids. For example, you want to compare prices on the same kitchen sink model numbers to see which contractor is charging a higher profit margin. Also, make sure the contractor has included his fees in his estimate. He may or may not break out his margin as a separate line item, but you need to know there won't be any surprise additions to the costs down the line.

Many contractors substitute an *allowance* for unspecified materials. So, for example, if you haven't picked your bathroom fixtures, one contractor may offer you a $10,000 allowance. This amount may seem cheaper than a different contractor's $15,000 estimate, but you have no way of knowing if the quality is comparable or who is taking a higher margin. Also, availability and the difficulty of installing certain materials can impact the time and, subsequently, the cost of the project. The surest way to make sure that all bids are assuming the same materials and labor necessary is to specify all the materials required for the project *before* you put the plans and specs out to bid. Having all these decisions clearly made in the beginning helps to avoid ugly misunderstandings and surprises later in the project.

Evaluating a contractor's 3 Cs — Cost, craftsmanship, and compatibility

Do you want a quick and easy guide to help you pick the right contractor? (Yes, of course you do, because you're reading a *For Dummies* book.) Kevin has been advising his clients for years to make their choice by examining the 3 Cs: Cost, craftsmanship, and compatibility.

✔ **Cost:** This one is obvious and, although important, it probably weighs the least in your decision for picking a contractor. The cost comparison becomes plain when the contractors return the bids. If you have handed them complete, detailed plans and specs, you need a clear picture who is working on the lowest margin or who has access to the lowest cost labor and materials.

✔ **Craftsmanship:** This part is important in the long term. You want to know that the house is built well and will offer you decades of enjoyable living. Don't confuse style and design with craftsmanship. A house can have a horrible floor plan with fantastic artisan work.

What's the best way to check on work quality? Look at other houses that the contractors built. Ask the contractor for a complete address list of homes he has built in addition to a list of references. Make sure you look at houses built ten years ago as well as newer ones. Remember, just as a car with 50,000 miles drives much differently than a new one, an older home lives much differently than a new one. Don't forget to talk to the people living in the houses to find out what problems, if any, they have experienced with their homes. Don't be afraid to knock on the door of homes built by the contractor that weren't on the reference list.

Also, don't forget to ask the contractor about his workmanship warrantees. Warrantees usually last for ten years, but can vary. See Chapter 15 for more on contractor warrantees.

✔ **Compatibility:** This "C" is the most important aspect to consider, and yet the most difficult to identify. The hard part is first assessing who you are so you can pick the contractor that will work best with your style. For example, if you're a micromanager that plans on being involved in every aspect of the project, then you'll constantly butt heads with a contractor that also likes to micromanage his projects. You need a more relaxed contractor willing to let you make the decisions or second-guess his work. On the other hand, if you don't have the time to be involved in the management, then someone with a relaxed attitude probably won't give you the sense of security you need — particularly if he is waiting for you every day to make decisions. Figure out what sort of experience is going to make you happiest and find the contractor that can meet your needs.

References from people you know are great, but people have different tastes and perspectives on quality and experience. Ask specific, open-ended questions about the experience that can paint you a clear picture of the contractor's personality. Just hearing that "It was a good experience" doesn't tell you much without knowing the reasons why.

Using expert interviewing techniques

If you haven't hired project managers before, then hiring a contractor will be a completely new experience for you. Generally, the contractor is trying to sell you on himself for the job because that is his business. At the same time, you're trying to sell the contractor into taking your project because the price is right and you like his work or maybe the market is really busy right now. With everyone so anxious to get going, the important issues, such as compatibility, can get passed over — leading to tense problems and unmet expectations later in the project. You need to make sure you not only ask the right questions, but also use effective methods to get the answers you want. The following list includes some tried-and-true interviewing techniques that are guaranteed to help you find the best contractor for your job:

- **Let the contractor do the talking.** If you're doing more than 25 percent of the talking, then you're the one being interviewed. Let the contractor explain to you why he wants this job. Have a standard list of questions for each interview that allows the contractor to tell you what he's like to work with and what services he provides. (We provide just such a list on the Cheat Sheet at the front of this book.)

- **Let the contractor tell you what he wants.** If you ask "yes-or-no" questions about his style and needs, he'll surely try to answer based on what he thinks you're looking for. Instead, ask him open-ended questions about his ideal project and the worst project he has ever had. Delve into details about his likes and dislikes. You'll be working with him for at least six months on this project, and nothing is worse than an unhappy contractor.

- **Give him problem scenarios.** Anyone can run a project that goes perfectly smooth all the time. You want to know how he deals with problem situations. Create stories of nightmare situations and ask him how he would handle them. For example, what if the framer and the plumber get into an argument, and one walks off the job before the job is finished? If he quakes in his boots with the questions, you'll know he doesn't have the strength to manage your project.

This project is your home, and your lifesavings are at stake. Compromise is okay to a point, but you need to have a good building experience to enjoy your home fully when you finally move in. Don't assume that contractors are all the same. Pick the one that makes you feel comfortable and secure and then communicate, communicate, communicate.

Identifying Insurance Issues

If you arrange outside financing for your project, the lender requires insurance to cover several issues, including at minimum

- Liability
- Workers' compensation
- Course of construction

Even if you finance the project 100 percent out of your own pocket, you still want to protect yourself. This section outlines the different insurance policies necessary to protect your project and yourself. Don't wait to discuss these policies with your insurance agent. Not all carriers have these policies available, so you may need extra time to find a carrier and shop for the best price.

Liability policy

Lenders require *liability insurance,* which protects you against someone getting hurt on the property or by the actions of somebody working on the property, to be carried by either you or the contractor. If the contractor carries the policy, the policy needs to meet the following criteria to satisfy most lenders:

- It must be in the form of a comprehensive general policy for $1 million or the loan amount, whichever is greater, or be a policy including broad-form liability endorsement.
- The contractor must be named as the insured.
- You and the lender must be named as additional insured.
- The property address must be included on the certificate.

You may want to carry this policy yourself because in some states it can be expensive for a contractor to get a liability policy if he doesn't already have one. If you carry the policy, the cost will be added to your budget and considered for financing as discussed in Chapter 8. Note that policies vary in cost by state and the size of your home. If you're carrying the policy or are an owner-builder, then you can get the liability policy with a few changes:

- ✔ The contractor doesn't need to be named.
- ✔ You're named as the insured.
- ✔ The amounts change to $500,000 for each occurrence, extended to both property and personal injury or the loan amount, whichever is greater.

Workers' compensation

Ordinarily, the contractor carries a workers' compensation policy if she has employees. You can ask to see the certificate because the lender will as well. In many cases, however, the contractor doesn't have her own employees and hires her labor and subs as independent contractors. If so, then these independent contractors fall under your liability policy in case of an accident. Lenders usually allow for this situation by having you and the contractor sign a waiver so the lender isn't held liable for any workers' compensation violations.

Course of construction policy

This policy protects you in case of theft, fire, weather, or other damage to the house while it's being built. So if, for example, your plumber drops his torch and burns down the framing, this policy pays for the cost of rebuilding the house to its previous condition before the damage. Lenders absolutely require this policy to be in place before they fund loans. The cost of this policy varies depending on your state and the size of your home. The following criteria needs to be included in a course of construction policy:

- ✔ Coverage must be in an amount equal to the estimated replacement value of the improvements to be built or the loan amount, whichever is lower. Guaranteed replacement is usually acceptable instead of a specific dollar amount.
- ✔ The borrower is the named insured.
- ✔ The lender is named as the "mortgage and certificate holder."
- ✔ There is a 438BFU Lender's Loss Payable Endorsement naming the lender as the "Loss Payee."
- ✔ The property address and/or legal description is listed on the insurance certificate.
- ✔ The maturity date on the insurance is at least one day beyond the end of the construction loan term.

Managing Your Expectations

One of Kevin's favorite contractors earned the "favorite" title because he is the biggest pessimist around. The problem with being optimistic with a construction project is that you're constantly disappointed. Materials don't show up on time or the sub takes longer than you thought to finish; you name it, it probably will happen. A custom home project is a journey with a life of its own. You can follow along and guide it, but you can tightly control only a relatively small portion of it. The best approach is to be well prepared, relax, and enjoy the ride.

Planning a timeline — A custom home is forever (So what's the hurry?)

In Chapter 1, we show you a list of all the actions and people involved in the custom home process. The actual time frames for this process can vary based upon the size and scope of the project as well as the city or town you are building in.

In order to make the planning process go as smoothly as possible, talk to a couple of architects and contractors early to get a sense of how long everything normally takes in your area. Write down each step and the expected time frame. Now fold up that paper and stick it in your file. This paper is a guideline for you to refer to now and again only as a reference. You can fully expect for some tasks to take longer than you estimated.

A house is built to last decades. You don't want to rush a complex process, such as design or construction, to the point where corners are being cut. An extra month or so may cause inconvenience and may even cost a little money, but in the end you'll have a better-constructed home and you'll have forgotten about that delay after you've lived in the house for a year or two. That being said, you definitely need to monitor the schedule and time frames. Pay special attention to deadlines associated with the building department and your lender. But use your best judgment and apply pressure only when it's necessary. Your project is a working environment filled with plenty of stress. Overreaction and constant pressure on the workers can make them less likely to help your project move along smoothly.

Making hard choices — What you can (and can't) afford

Most custom home projects go over their original budget — some more than others. You really have no way of knowing how much your project will cost until it's actually finished. Until then, you have to constantly make choices to adjust your budget's expenses. Take a breather when making these decisions. Think about whether that gold-plated faucet really makes an important difference in your lifestyle. Sleep on it for a night or two. Find creative ways to make the home something you will enjoy while spending less money. We share some of our favorite ideas for saving dough in Chapter 19.

Give yourself a break. Design a house that doesn't stretch you to your limits. That way if you're coming in on budget, you can choose to upgrade and splurge here and there without worry. (See Chapter 5 for more tips on saving money when designing your house.)

Patience — Not everything must be perfect right away

Many craftsman and artisans participate in the building of your custom home. If you've carefully hired everyone, you need to put your trust in these skilled workers. Many processes are multistage. Unfortunately, some people will make mistakes or leave some items partially finished. Don't panic; this process is normal. Your contractor and the subs can fix or replace most everything during the building process, and your contractor and subs do double-check their work as they go.

We give you spot-checks in the chapters in Part III of this book. Use them as a guide to check on everything along the way. Write down any concerns and problems you see. Rather than pointing out everything every time you have a concern, talk with your contractor and set up a meeting with the appropriate sub to share your findings. Doing so allows the subs to do their job with a minimum amount of pressure. You may find they had already scheduled the repair.

Keeping an eye on everything — Cameras on the property

If you absolutely need to know what is happening with the property 24 hours a day, you may consider putting up cameras and broadcasting the signal to the Internet. Doing so is perfect for all you gadget geeks building custom homes. Internet sites such as www.x10.com/cameras sell cameras and software that allow you to watch the action at your build site from any computer connected to the Internet. You can expect to pay a couple of hundred dollars for all the equipment you need to watch your project 24 hours a day. You may find it isn't much different than watching paint dry, but what's a little boredom where peace of mind is concerned?

Making the process fun

You're sure to have your share of frustrations and problems before your custom home is finished. Managing your patience and your temper is difficult when so many things are beyond your control and so much is at stake. To help you enjoy the lighter side of the process when everything is looking a little dim, we give your our top-10 list of ways to make your process extra fun:

- Enjoy a mud football game before the foundation goes in.
- Throw a block party at every completed stage.
- Have an office gambling pool on completion dates.
- Make matching T-shirts for all the workers on the project.
- Create art pieces with scrap lumber and supplies.
- Compile a construction hunk calendar.
- Give out a worker-of-the-month trophy.
- Put your family's handprints in the cement.
- During framing, fly paper airplanes off the second story.
- Make a photo album/scrapbook documenting the entire process.

Chapter 3

The Land Grab: Selecting the Perfect Site

You can't have a house unless you have somewhere to put it. Unlike the pioneers of old, however, individuals today can't just walk up and squat on whatever piece of land that catches their eye. If you're interested in some land, you have to research, explore, negotiate for, and ultimately purchase a parcel that you can call home.

In this chapter, we walk you through the entire process of searching for and finding the perfect site. We provide you with the very best evaluation tips, and even discuss buying a house in poor condition that you can tear down and rebuild. Finally, we consider financing options. You have to pay for that dream property, after all!

Knowing the Difference between "Land" and a "Lot"

Differentiating between some land and a lot may seem like an easy distinction. But not so fast. The two are in fact quite different. All lots can be considered land but not all pieces of land can be called lots. Are you scratching your head? If so, read on . . .

✔ A *lot* or *finished lot* is a piece of property that is ready for building a house. It may or may not have all the utilities (gas, electric, water, sewer, telephone, and so on) at the site, however, they usually aren't far away.

✔ *Land* is a catchall term that people in the construction industry use for any piece of property without a finished, habitable structure. Land can be commercial, residential, or agricultural. *Raw land* or *undeveloped land* is terminology that most people in the construction industry usually use, referring to land that isn't ready for building.

If you're anxious to get your new home built or if your finances are limited, then plan to buy a finished lot, not raw land. You can find a good loan more easily with a lot, you'll likely pay a lower down payment, and you'll have to spend less to prepare the property for building — all resulting in major savings in both time and money.

Raw land can be more difficult or more expensive to finance because it usually requires additional work (often, *significant* additional work), such as putting in roads and utilities, before building can begin. Fewer buyers are willing to put the time and effort into this type of property, which makes it less marketable than finished lots. As a result, fewer banks will be willing to lend you the money you need — you may have to find private financing or encourage the seller to loan you the money (referred to as having the seller *carry back paper*). If you do find a lender, you'll probably need to make a larger down payment. In most communities, noninstitutional individuals invest money in real estate — banks usually know these people. (See "Using private or hard money" later in this chapter for more information.)

Also, if you buy raw land, plan to allow yourself more time to complete the entire project because preparing raw land into a finished lot — getting approvals and permits, building roads, extending utilities to the lot, drilling wells, if necessary, and more — can take months or even years. Make sure you consider the extra time when discussing the term of your lot loan with the seller or private lender.

Location, Location, Location — Refining Your Lot-Buying Needs

Of course you don't want to buy just any lot. You want a lot that meets with your desires and will retain its value, if not appreciate.

Several criteria can affect the sales price of different properties both today and when your house is finished. Picking the right lot is just like picking a finished house. Factors such as location and amenities can make a lot undesirable. If a lot is undesirable to you, it may be undesirable to others as well, which can have an effect on the marketability of the property and, ultimately, the resale value. Consider many factors when deciding on the right location for your lot.

The following sections contain some questions to ask yourself. When answering them, take into account not only your own lifestyle, but also the factors that will impact your ability to resell the finished house. Remember, use this checklist when searching for the right neighborhood as well as when evaluating a prospective lot for sale.

Lifestyle

You need to decide how you want to live in your home — these elements are a matter of personal taste. Many of the lifestyle items in the following list may be important to some of you and unimportant to others:

- Should the lot be in an urban, suburban, or rural area?
- Should the lot be flat or sloped?
- Should the lot have much usable land?
- Will the lot require significant ongoing maintenance?
- Should the lot be in close proximity to the neighbors?
- Should it afford privacy?
- Should it be sunny or shady?
- Should the lot have natural vegetation?
- Should the lot have available on-street parking?

Marketability

The list that follows has a variety of factors that impact value, and you need to use these factors when evaluating a lot to purchase. Keep in mind that you may have to sacrifice some of these factors, however, to meet your budget. Make sure you talk to a real estate professional to understand the market demand of a particular lot based on these issues.

- Is the lot on a busy street?
- What kind of view does the lot have?
- Does it have waterfront access?
- What is the proximity to power lines?
- Is it next to or near commercial buildings or apartments?
- What's the noise factor? How close is it to planes, trains, and automobiles?

   ✔ Is it on or near an earthquake fault or in a flood plain?

   ✔ Are there good schools in the area?

   ✔ What type of power, water, and sewage is available?

   ✔ What does city planning have in mind for this neighborhood?

Finding a Lot

Acquiring the right lot may be the biggest challenge you'll face in the custom home–building process. Property values have been steadily increasing in most places since World War II, and not every piece of land will suit your particular needs or be cost effective for building. Finding the right lot isn't easy, and you need to plan for a long hunt. Some helpful resources are available, but not as many as you may think. Finding the right lot requires sleuthing and persistence. Your best resources are using the Internet, utilizing experienced real estate agents, and taking your own initiative.

Surfing for turf

Using the Internet is a great way to start looking for your lot. The following Web sites are our three favorite picks for discovering your dream lot:

   ✔ www.lotfinders.com: This site provides educational information on purchasing lots and connections to real estate agents that sell lots. Financing information is also available.

   ✔ www.realtor.com: Owned by the National Association of Realtors, this site can show you every piece of land listed on the real estate bible: the Multiple Listing Service (MLS).

   ✔ www.land.net: This site has listings for large parcels of agricultural and residential land as well as individual lots.

Engaging a real estate agent/lot specialist

Finding a real estate agent that specializes in lots can be difficult and frustrating. Most agents don't want to spend the time with a buyer only to sell a piece of property at a fraction of the price (and a fraction of the commission) of a completed home; however, a few agents have made dealing with lots their specialty. Check to see if the real estate agent you're working with understands the issues associated with building, planning, and zoning that we reference in Chapter 5 as well as in this chapter. If your agent doesn't readily answer your questions and seems unsure, if you feel like you know more about these topics from reading this book than she does, or if she brushes off or blows over your questions entirely, you may want to look for someone more experienced in lot sales. If your agent seems knowledgeable in lot-only sales, she can help you assess whether any potential problems associated with building on a particular lot may surface.

Even if you can't find an agent who focuses on lot sales, working with an experienced real estate agent at some point still may be in your best interest. Even though you may have to find the lot yourself, you still need help with the negotiations and transaction management. If the property is listed in the MLS, the seller is paying a commission of 3 to 5 percent to each agent anyway, so you may as well have an agent representing your best interest. Otherwise, the listing agent gets the entire commission just for representing the seller. (Check out Wiley Publishing's *Home Buying For Dummies,* 2nd edition, by Eric Tyson and Ray Brown for more info on real estate agents and their duties and commissions structure.)

Doing the legwork on your own

If you want to find your perfect lot, you may have to do some detective work. Grab your sleuthing equipment and get started.

If you're looking for large tracts of custom lots, try the outlying areas of your city. Custom home developments often advertise in the supermarket real estate magazines, and new golf courses can make for a hotbed of lot subdivisions. Sometimes you can find the right lot simply by spending your Sundays driving through neighborhoods under development and looking for signs.

Finding a lot when there isn't one

If you have searched online, worked with an agent, driven the neighborhoods, and still haven't found your dream lot for sale, you may need to take a more aggressive approach. If you find any piece of land you like, then contact the owner and make an offer! It doesn't matter that it isn't listed for sale right now. Remember, everything is for sale; it's only a matter of price. Information on who owns any piece of land is part of the county public record. Through a real estate agent, mortgage broker, or title company, you can request the address and phone number of the owner of any piece of land. If your real estate agent is managing your part of the transaction for a commission, she should be thrilled to help you with the negotiations and the closing process.

If you're willing to look at many pieces of land, you can take the shotgun approach. Have your agent get you a mailing list of all the lot owners in the area and send them all personal letters explaining your burning desire to own property and build in the neighborhood. Perhaps someone will consider selling his property and, if not, he may know someone in the neighborhood who may.

Too bad it's not 1889!

In 1889, Oklahoma was emerging as a wealth of opportunity for settlers looking for land. Oklahoma Station and Guthrie Station were two promising railroad outposts destined for urban development. In one of the strangest and chaotic stories of land acquisition, the government created an exciting and unprecedented process for claiming land — a race! Rules were posted allowing people to gather at the nearby Arkansas and Texas borders ready to run, ride, and walk to their desired parcel of land on April 22, 1889. Upon arriving at their parcel, they would claim it by staking a flag and filing a claim form. People already in the territory ("Legal Sooners" as they were called) would cheat by staking their flags early even though prohibited by law. Thousands raced that day and made Oklahoma land their home. It was exhilarating and brutal. "It is an astonishing thing," the *New York Herald* observed on the eve of the opening, "that men will fight harder for $500 worth of land than they will for $10,000 in money."

Evaluating a Particular Lot — The True Value of Dirt

Although it may be true that the value of something is based upon what someone is willing to pay for it, land value has other factors to consider when determining its ultimate usefulness for building a custom home. You need to take into consideration the cost of getting the land ready for the build. You also must factor in limitations on the size of the home.

When determining a particular piece of property's value, a lender's appraiser looks at other comparable land sales in the area to create a number for the lender's purposes. Keep in mind, however, that this appraisal doesn't guarantee your ability to resell the land for the same price. Nor does it mean that your custom home budget will be able to absorb the price of the land. Consider all the factors associated with your build — especially financing — before purchasing a lot. Chapters 8 and 9 can help with understanding how lenders evaluate your land in relationship to the entire custom home project.

Examining amenities and utilities

The relationship between utilities and your lot can have a significant impact on the lot's value. A lot requiring a septic system can add costs (which may decrease its value). The need to drill a well or to add offsite additions, such as sidewalks and parkways, can also negatively impact a lot's value. Be sure to explore the cost of installing utilities and amenities before you buy any lot.

This section contains some "due diligence" items for you to consider when figuring the ultimate "cost" of your lot. Ask these questions of the seller and your agent, or research them with the appropriate county or city agency. Then create a list of all possible costs to prepare for your budget and estimate how long it will take to work through the permitting or approval process. Here is short list of questions to ask:

- Does your lot require gravel, asphalt, or concrete for its driveway?
- Will you need to install sidewalks?
- Will you need to install street lighting?
- Will the lot require extensive earth moving or a special foundation?
- How far is the electricity from the build site, and how much will it cost to extend it to your property?
- What is the cost of connecting to the sewer or installing a septic system?
- Will the property support a required septic system?
- What is the cost of connecting to water or installing a well?

Many people get so caught up in the dream of building a house that they sell themselves into impractical situations. Keep your cool and do your homework. Buying a piece of land without researching all the issues and costs can leave you with a crushed dream *and* a useless piece of land.

Zoning in on zoning's limitations

Nearly every city and county attaches building restrictions to land when the land is first put into development. This is called *zoning*. Zoning determines many factors you must consider, including

- The type of building you can put on the land
- The lot's minimum size
- The number of dwellings or units you can build on the lot

Finding out and understanding the zoning of a particular lot before you decide to buy is important. The information is readily available at your friendly, local city hall or county government building; your local government also has a guide that can tell you what each zoning designation means.

Most urban and suburban lots for houses are zoned residential R-1. If a zoning designation has a higher number, you can build more units on it (for example, a duplex or triplex) based on whatever number is designated (R-2, R-3, and so on). Some commercial lots can be used for residential, but the rules vary among municipalities. Most large, rural properties are designated RA for Residential Agricultural. When the zoning has a number such as RA-5, the designation means your lot must be at least 5 acres in size. This information is important to know in case you want to split the property into more than one lot. If you have 9 acres with RA-5 zoning, you can't divide the lot in two. (Remember, each lot needs to be at least 5 acres.) You could split a 10-acre parcel in two, providing it meets all other county zoning guidelines.

Don't take today's zoning for granted — the local government can zone to meet with the needs of the community on almost a moment's notice and without your agreement or even advance knowledge. Although this change happens mostly in rural areas, check with the local government to see if any zoning questions for your area are scheduled. While you're at it, be sure to ask if any special tax assessments will affect the property. These assessments can add to your expenses as well.

Even though you can in theory build a single-family residence on a lot zoned for commercial or apartment use, you may want to take the zoning into long-term consideration. Often, a property owner may destroy a small house in a commercial area and build apartments or office buildings, which could change the neighborhood's tone for the worse (think noisy neighbors and cars parked everywhere), reducing your home's value.

Zoning only affects the general plans for the property. More detailed restrictions for building may also be stated in the covenants, conditions, and restrictions (CC&Rs) that are recorded on the property as well as design review guidelines. We talk more about CC&Rs and design review guidelines in Chapter 5. Ask the homeowners' association (HOA), your real estate agent, or title company for a set of CC&Rs and design review guidelines to find out about other restrictions, such as setback limits, style limitations, and height limits, before you agree to purchase any lot.

Understanding setbacks and footprints

Setbacks determine how far from the edges of your lot you must build. They're generally determined by the property zoning restrictions and the CC&Rs. The side setbacks are generally closer to the property line than the front and rear setbacks, but not always. This information is crucial for figuring out where you can place the house on the lot. Many neighborhoods like to keep the houses uniform, so look to see how close neighbors' houses are to the street and each other to get a feel for the setbacks. In urban areas, the side setbacks may be as small as only a few feet (Peter's old house in San Diego had a side setback of only 3 feet — barely enough room to wheel his trashcan through to the street every Monday night). Rural areas can require larger setbacks from the street or other houses, impacting curb appeal.

Setbacks also apply to outbuildings, not just to the main house. On larger lots, you may be considering the possibility of guest homes, workshops, and pool houses within the buildable area. These buildings all have to be within a certain area defined by the setbacks.

The *footprint* of the home is the building's outer perimeter and how it sits upon the lot, taking into account the setbacks (see Figure 3-1). When looking at a lot, try to imagine in rough design how positioning the footprint can take advantage of the following:

✔ Drainage

✔ Noise

✔ Sunlight

✔ Topography

✔ Views

✔ Wind

Figure 3-1:
Setbacks determine the placement of your footprint, ultimately restricting your home's size.

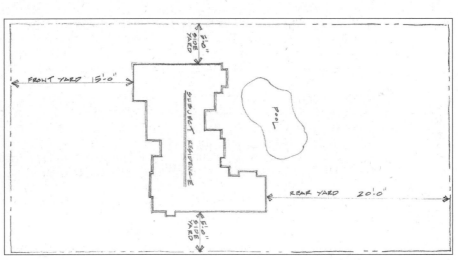

Courtesy of Tecta Associates Architects, San Francisco

The combination of footprint and setbacks may dictate whether your house is one story or more. Significant side setbacks can force you to design a smaller footprint home, which can make a difference in estimating building costs. A smaller building area may mean building a second floor is essential. Building a second story can work to your advantage, however, because a two-story house can be less expensive to build than a one-story home. (For example, building a two-story house with 4,000 square feet is cheaper than building a one-story house with the same square feet. The two-story house requires less excavation, foundation, and roof work than the one-story house.)

Height restrictions spelled out in the CC&Rs have an impact on your house's size. If your setbacks force a smaller footprint, then your height restriction will limit the size of house you can build. Have your real estate agent obtain this information for you *before* you purchase the lot to make sure it meets your needs. You can also contact the local planning department to access the information yourself.

Size matters — Assessing the land's value with the house

Even though much of the planning for your new home is considered in the design phase that we discuss in Chapter 5, considering the planned size of your house when purchasing the lot is important. You can't build any house you want on any lot. You'll probably face restrictions set by the city, county, and sometimes the market. Doing your homework on the limitations of what can be built on a lot can keep you from making costly mistakes.

How do you know for sure that the land you want to buy is going to fit within your budget? You must consider it in the context of the budget for your entire custom home–building project. (See Chapter 2 for information on budgeting.) Start by researching the sale prices of houses in the area. If homes in the neighborhood are selling for $500,000, then buying a comparable lot for $375,000 isn't likely to leave you with financial room to build.

Still not sure what to do? Here's Kevin's quick-and-dirty four-step process for getting a rough idea if a lot is too expensive:

1. **Contact a real estate agent and get a list of properties that have sold in the area during the last year.**

   Make sure the list includes the square footage and room count of each home.

2. **Pick the three sale prices that are most similar to the house in square footage and room count that you want to build and average the sales prices.**

Most lenders use at least three comparable sales to establish an appraised value, so this number helps you evaluate the property from the lender's perspective. If no home sales are similar to your desired home in square footage and room count, then you may need to reassess your design or your choice of neighborhood.

3. **Subtract the land price from the average sales price and divide by the square footage you want to build.**

   Doing so gives you a rough number for *dollars per square foot*. We explain dollars per square foot in detail in Chapter 2.

4. **Call three contractors in the area and ask if they can build for the dollars per square foot number you established in Step 3.**

   The contractors you call at this point can be referrals or out of the phone book. Where you locate them doesn't really matter because you may not use them for your project anyway, you simply want a rough survey. For more information on contractor selection, check out Chapter 2.

Taking these steps can give you a rough idea if you're even close. This method, of course, does have many variables and unanswered questions, such as the selection of your fixtures and materials and foundational needs. However, if all three contractors you contact are rolling on the floor with laughter, then you're probably looking at an overpriced lot.

A tale of two lot buyers — How square footage impacts value

Kevin relates this experience from a custom home project he financed in northern California.

Ten newly divided lots were being sold for $200,000 in an established neighborhood. Frank Smith looked at one of the lots and was concerned because he thought the lot was too expensive. He was absolutely right. Mark Jones came in the same day and looked at the same lot and quickly determined the lot was a great deal. He was also absolutely right.

How can they both be right if they're talking about the same lot? The decision is all about the square footage. Frank wanted to build a 2,500-square-foot house, which was comparable to other houses in the neighborhood selling for $400,000. The total cost of the house including the land penciled out to $450,000, making the project too expensive to build on this lot. Frank wouldn't be able to borrow enough money to build his house.

Mark's house was going to be 4,500 square feet. His cost per square foot was the same as Frank's so his total cost for the project including land would be

$650,000. Houses of *this* size in the same neighborhood were selling for $700,000, allowing Mark a $50,000 profit on his house, which allowed Mark to borrow plenty of money to build his house.

Your real estate agent can show you the sales from the last year to evaluate the optimal house for your neighborhood, which can help you make sound choices for maximum value and the best financing.

Dealing with a Tear-Down Property

Buying vacant land isn't the only option for locating a custom home. Many people opt to buy a small or dilapidated house and tear it down or add to it significantly. Doing so can be an excellent way to move into a new home but still have all the benefits of an established neighborhood.

Accounting for demolition costs

One immediate cost benefit to a tear down is the fact that all the utilities are already located on the property and connected, which can be a great cost savings, but you still have an old house connected to them. Contact a demolition contractor and get an estimate for the demolition and clearing of the property before you sign on the dotted line.

Demolishing an older home may have additional costs due to the removal of hazardous waste such as asbestos (often located in roof, ceilings, or siding). Professionals need to remove asbestos, which can be costly.

Leaving one or two walls of the old home standing may be cost effective and worthwhile. Why not tear everything down and start fresh? Because many cities or counties offer a lower property tax rate for a remodel versus new construction. Depending on your community's rules and regulations, leaving one or two existing walls establishes the project as a remodel and can save you thousands of dollars a year.

Assessing neighborhood tolerance

When you rebuild in an established neighborhood, you want to make sure the house conforms to the neighborhood. This action is critical to ensure a marketable property in the future. If you decide to locate your 3,000-square-foot home in a tract of 1,500-square-foot homes, you'll never get the value you want out of your home. The neighborhood needs to support the value. Underbuilding can also be a problem because people looking to buy in the neighborhood will expect a comparably sized house.

Most established neighborhoods require community approval of any major remodels. The local planning department generally notifies the community, and residents have the opportunity to voice their opinions — good or bad. Some areas have a specific design review committee that has a significant say in what you can and can't build.

Do your research and build a house that fits within the neighborhood tolerance or find another neighborhood that has houses closer to your desires and needs. You can find more information on the design review process in Chapter 5.

Financing pros and cons

You may also encounter financing advantages if you buy a house that needs to be remodeled instead of buying a lot. If the house is still structurally sound, you may be able to buy the house with little or no down payment, while a house with structural problems can create financing problems. Most banks don't lend money on a house that is in functional disrepair (if you default on your loan, the last thing the bank wants is to be the proud owner of a hunk of junk). Buying such houses requires a lot of cash or private money, which can be expensive. Even then, a large down payment is necessary.

Small houses on large parcels of land can also be a problem in neighborhoods where bigger houses are now being built. Lenders want to finance houses for living. House lenders want a property that conforms to the neighborhood, even if the borrower is well qualified. Many people, when faced with unwilling house lenders, look to lenders that lend on land only. If the property has any structure on it at all, most land lenders won't provide a land loan. Talk to your real estate agent and loan officer about the ability to lend on a particular property before you make an offer. Remember, the finished cost estimates of your project need to account for the purchase cost as well as the demolition cost in order to be realistic.

If you have found a structurally unsound house and a willing seller, you may be able to buy the home with a construction loan, which will finance the purchase of the property and the construction together based on the future completed value of the property. Chapters 8 and 9 explain in detail the construction loan process and guidelines. You can apply the same construction loan information in this book as if you were working with a vacant lot. Simply have your seller agree to a long escrow that allows you the necessary time to design and permit your new house. Most construction lenders can fund the purchase as part of the construction loan provided all their other guidelines are met.

Buying Your Land

After you find your dream lot, you need to buy it. Whether you pay cash (a big no-no — Chapter 7 explains why) or finance, you can't just write the owner a check and ask for a receipt. Okay, you could, but you need to follow certain steps and make specific choices if you want to get the most for your money.

Understanding the purchase process

If you have ever purchased a home, then you're at least somewhat familiar with the process of buying real estate. Buying a lot is just like buying a house, but without all the house stuff. You can find a detailed explanation of the purchase process and the players involved in Kevin's book *What the Banks Won't Tell You; How to Get the Most Out of Your Mortgage* (Grady Parsons Publishing), available at www.stratfordfinancial.com. For beginners, here is a general step-by-step look at the process.

Step 1: Determine your offering terms and price

Using the information in this chapter about evaluating a property, you need to determine a price in your mind of what you're willing to pay for the property. Your real estate agent can give you additional information on the local market for land and examples of finished houses in the area. If you aren't using an agent, you need to access this information through the local title company. You also must decide on the length of time until closing as well as any other necessary terms, such as your desire for the owner to loan you the money in what is called a *seller carry back*. This term means the seller, after transferring the property, retains a note or loan that you must pay him at an agreed time with interest. We talk more about this topic in the "Finding other land loan alternatives" section later in this chapter.

Step 2: Present the offer and negotiate

You or your real estate agent has to complete an offer form accompanied with some sort of good faith deposit check from you. The deposit amount can be anywhere from 1 to 3 percent of the purchase price. The check is deposited in an escrow account upon acceptance of the offer and held in escrow until you close the transaction. The offer and copy of the check are then presented to the seller and negotiations begin.

You and the seller can negotiate the deal through subsequent documents called *counter-offers.* The seller's willingness to negotiate is relative to how hot the real estate market is at the time. A hot market usually translates to less willingness to negotiate. The seller isn't obligated to accept any of your terms and can always choose to say no. Of course, you can always choose to walk away. Ideally, you'll each give in a little and come up with a workable compromise. As Kevin's mom, a 30-year real estate agent veteran, said, "An offer is an opening of a conversation." The less emotions in the conversation, the easier the negotiations proceed.

If the property is listed with a real estate agent, that agent doesn't represent your interests. Make sure you bring your own agent to represent you as the buyer. Doing so doesn't cost you any more because the seller pays the commission. If no real estate agent is involved, consult one for guidance. She may help you with the paperwork for a small fee. You can always check with an attorney as well, but an attorney generally charges more and sometimes makes the transaction more complicated than it needs to be.

Step 3: Make an application with your lender

We explain various financing options in the next section. After you have determined which lending approach is right for you, you'll fill out a loan application with the lender of your choice. The lender orders an appraisal from a certified appraiser. You'll probably have to pay for this appraisal upfront. The cost can vary depending upon location and the lot's value, but usually it will cost between $300 and $600. You're entitled to a copy of this appraisal, which will be based upon comparable lots in the area that have sold in the last six months to a year. The original appraisal goes to the lender along with the application and any credit documentation you provide, such as bank statements, W-2s, and tax returns.

At this time, your lender will give you a good faith estimate (GFE) of all the closing costs associated with your loan and the purchase transaction. The costs vary based upon the loan amount and type of loan, but you can anticipate a range of 3 to 6 percent of the purchase price as an estimate of all the costs involved.

You can save the upfront cash for closing costs by offering to increase the price of the lot by 3 percent and then asking the seller to credit you 3 percent of the purchase price for nonrecurring closing costs. Most lenders accept this agreement as long as the appraiser mentions that it has no effect on the value. Financing the closing costs in this way leaves your cash available for other important costs along the way.

Step 4: Open escrow with an escrow company or attorney

The term *escrow* means depositing money and property with a neutral third party to be disbursed upon completion of all terms of a related agreement. Each state has its own process for escrow procedure. Some states use attorneys to act as the escrow agent and others use title insurance companies. If you're in a *wet funding* state, you, the seller, and the escrow agent all pick a day to meet and sign the paperwork at one time. In a *dry funding* state, such as California, you each execute the paperwork on your own or with a notary public over several days before the actual closing date.

The escrow period can be anywhere from 30 days to 6 months depending upon the needs and negotiations of you and your seller. Your real estate agent helps guide you through the closing process. If you don't have an agent, then the escrow agent will be your best guide. You can find good escrow agents through referrals from real estate agents, loan officers, and friends. Most are well trained, so unless your transaction is extremely complicated, you should be able to go with someone you simply find personable.

Step 5: Do your due diligence

While you're in escrow, you have a chance to complete any research on the property that couldn't be done before the offer. In a hot market, you might not have had much time to research the property before putting in an offer. As soon as you have a chance, make sure that the property meets your needs relative to size and value as outlined in this chapter. If you have concerns about the land, such as building restrictions or guidelines, you may want to add contingencies to the offer that allow you to pull out of the transaction if your research results are unfavorable.

Step 6: Execute the paperwork and bring in the money

As soon as your loan is approved, you need to bring your cash to escrow and sign the loan documents. You can wire money into escrow or bring it in the form of a cashier's check. Bring your pen because you'll have as many as 100 documents to sign for the escrow and the loan.

Step 7: Close escrow and take title

The title insurance company provides you with a deed to the land and a policy insuring that it is yours. The title company pays the seller the money due and records any documents and deeds related to the transfer and new loan.

Using the bank

Most people finance their lot purchase through a local bank or nationwide lender. A good mortgage broker can help you determine who has the best programs to meet your needs. Local banks generally have more conservative criteria for loaning on land because they're heavily regulated. Large, publicly held lenders have the ability to offer creative and flexible loan programs because they mix the risk with other loans in their portfolio.

Not all mortgage brokers have experience with land and construction loans, so be picky. Don't trust the phone book on this one; ask around to find the experts. Try to find a loan officer that has 100 or more loans of these types under her belt. The best way to test loan officers is to see if they ask you more questions than you ask them. If they simply try to sell you on one type of loan without inquiring about your needs, then look for someone else to help with your loan needs.

Qualifying

The first thing a lender will ask you is whether you intend to buy the land for your own personal use. The lot-financing rates and terms for owner-occupied properties are much better than for investment properties. The lender looks to see if it makes sense for you to move to this property. If you're claiming it to be a second home, the lender will expect it to be in a resort type area or a city other than your primary residence. Buying the lot in a cheaper neighborhood on the other side of town from where you currently live will raise eyebrows.

The lender next assesses your qualification on the basis of your credit report, *liquid assets* (cash, stock, or other easily accessible forms of money), and your *debt-to-income ratio* (the amount of debt you carry in the form of loans and credit card balances versus your income). Your lender's approach to these issues is very similar to how it will underwrite your construction loan; see Chapter 9 for the specifics. To make its decision, the lender wants to see, at minimum, the following documentation:

- ✔ Appraisal
- ✔ Credit report
- ✔ Three months' bank statements
- ✔ Two years' W-2s and recent pay stub
- ✔ Two years' tax returns, if self employed

Banks may loan you a higher percentage of the purchase price based upon the quality of your other qualifications. Being able to show good credit and sufficient income can get you a loan for 90 percent of the purchase price.

If you have excellent credit, some institutional banks such as Washington Mutual and IndyMac offer "stated income" programs for purchasing lots. These loans don't require you to show any income documents (pay stubs or tax returns). Some lenders offer these loans up to 85 percent of the purchase price; however, the income you state must make sense relative to your job. A fast-food clerk supposedly making $150,000 a year isn't likely to get approved. You may need to show other documentation and meet other criteria necessary for these loans, so make sure you get the info from your loan officer before you apply. We give specific details on these types of loans in Chapter 8.

Many lot lenders are the same lenders that finance construction. You apply for a construction loan some months after you get the land. If the lender sees conflicting information regarding income or assets, the lender could turn you down. Therefore, understanding the requirements for lot loans and construction loans is critically important. (We explain construction loan requirements extensively in Chapter 9.) Lenders look in their files for any other loans they made to the same borrower. If the information is inconsistent, the lender simply rejects the loan.

Stated-income loans are notorious for creating consistency problems. Because a lot loan is usually a lower loan amount than a construction loan, less income is required to qualify. If you only state enough income to qualify for the lot loan, and then when you submit the construction loan application to the same lender you state a higher stated income amount, the lender will immediately deny the loan. Work with your loan officer to determine the qualifications necessary for the larger construction loan so you can represent the land loan file in a way that will match the construction loan needs. Chapters 8 and 9 can serve as a guide for that discussion.

One of the advantages of working with a mortgage broker is that she can act as a filter. By looking at your documentation, a good broker can determine which lender fits best with your project. Doing so keeps you from stabbing in the dark and providing too much information to the lender that may result in denial.

Picking a loan

The most important criteria for your loan is the loan's length of time. (We talk more about timing of lot loans in the section "Making sure the loan period is long enough" later in this chapter). Generally, your lot loan picks *you* based upon your qualifications. You may, however, need to choose between a *fixed-rate loan* or an *adjustable-rate loan.* Some lenders offer only fixed-rate loans where the interest rate stays the same for the loan's life. These rates are generally higher than adjustable-rate mortgages, which have interest rates that move with a particular monetary index such as government treasury bills.

Some people believe a fixed rate can save you money because it protects you from rising interest rates. But if you plan on building in the next few years, you'll be taking a construction loan that pays off the land loan. See Chapter 8 for details. Because you'll likely pay off the land loan soon with the construction loan, using a fixed-rate loan isn't likely to save you much money. Ultimately, you need to do the math and compare the various loan payment options over the length of time you anticipate before you start building.

Kevin's recommendation is to go with the largest amount of money you can borrow for the longest period of time with the lowest payment. Doing so gives you the most flexibility for moving into construction financing by keeping the maximum amount of cash in your pocket where you need it most.

Getting denied — What the banks won't finance

We can think of several situations that will eliminate conventional financing as an option. Some are based upon your own situation and some on the property. Here is a quick checklist:

- ✔ If your credit score is below 620 (see Chapter 9)
- ✔ If you have been late on your mortgage in the last 12 months
- ✔ If you're unemployed
- ✔ If you have no down payment
- ✔ If the property has existing buildings on it
- ✔ If the property is more than 50 acres (some banks allow only 20 acres)
- ✔ If the property has no electricity nearby
- ✔ If the property has no public access
- ✔ If the property has multiple parcels (some banks allow two)

If your property or qualifications fall into one of these categories, don't panic just yet. Other lending alternatives are available. Some may cost more money and be more restrictive than conventional lending, but they may be better than the thought of abandoning your project.

Finding other land loan alternatives

If you find the perfect lot and the bank thinks the property or your credit and income are less than perfect, you still may be able to buy it without paying all cash. Some lending alternatives are available if your property or credit doesn't meet the bank's guidelines.

Letting the owner carry the burden

One alternative way to finance a property is to have the property owner loan you the money or *carry back paper.* In this case, the seller acts as the lender and has you (the buyer) execute a note secured by the property for the amount he doesn't receive in cash through escrow. Not every owner will consider this option. If the owner carries back the entire amount of the purchase price, the seller can't owe any money on the property. Another way a seller can help is to carry back a second loan so that you can put down a smaller down payment. The primary lender must agree to this arrangement. In either case, the seller has the ability to foreclose if you miss payments. You need your real estate agent or an attorney to draft the note and security instrument to make sure everyone is properly protected.

When is a seller more willing to carry the financing himself? A seller may agree to carry paper if the property is hard to finance through banks or if he — for whatever reason — is anxious to sell. The seller may also want to defer the taxes due from the sale of the property; carrying paper allows the seller to pay taxes only as you pay off the loan. You gain no real advantage when the seller carries the financing unless the loan's terms are more favorable than any other lender offers you. Most construction lenders require the seller to be paid off when they fund the construction loan. Few institutional construction lenders allow a subordination of a seller carry.

Some sellers want a premium if they're going to carry paper. Furthermore, many sellers still want to check your financial wherewithal, so credit and income can still factor into their decision. Ultimately, you can negotiate the best deal with a seller if she is getting all the money expected from the escrow, so having her carry may not be the best route.

Using private or hard money

Hard money comes from private investors who specialize in making loans on real estate. Hard-money lenders generally aren't concerned with credit or income. They hope to make high-interest yields or make money by taking back your property through foreclosure and selling it at a profit. Typical hard-money runs a number of percentage points higher interest than the prevailing market rate, plus 5 percent of the loan amount in upfront fees called *points.* This high interest seems expensive, but if banks or owners won't give you a loan, then this choice may be better than not buying the lot at all. Because hard-money lenders like equity, they usually want as much as a 50 percent down payment.

Making sure the loan period is long enough

Lot loans come in a variety of lengths, but only a couple of banks offer them for more than five years. Your lot loan needs to be in place until the construction loan pays it off. Most projects can make it to the construction-loan phase within two to three years. If you think you're going to take a long time to design your home or that you'll need to save your money for a long time before beginning construction, then you may want to search for loans that last more than ten years.

How long it takes to begin the building process can vary wildly. The ultimate amount of time is based upon your local planning departments, how picky you are with your plans, how busy the current construction climate is, and many other factors. Figure out how long you think it will take and double it to be safe. Most of the delay factors will be beyond your control.

Stop! Don't pay off your lot yet!

Contractors, consumers, architects, and many others often tell you that you must pay off your lot before you get a construction loan. This is the biggest myth in the custom home–construction world. Actually paying off the lot isn't a good idea unless it's absolutely necessary. The following sections explain several good reasons to keep a loan on your land until you're ready to build.

You need cash on hand to fund your project

Buying your land is just the beginning of paying people in a construction project. The architect, the engineer, the well and septic people — and many others — need to be paid along the way. The permitting process can suck your cash as well. These people and processes can add up to tens of thousands of dollars. If you run out of money because you put all your hard-earned savings into your land, your new home can become a nightmare. Having cash in your pocket is your best protection for keeping your project moving along. Check out Chapter 7 for more on this subject.

Money put in is expensive to get out

Few lenders refinance a land loan; most only let you replace an existing loan. Rarely does a bank give you a loan where you're taking cash out of a piece of land. That means that after you put money into the land, it's gone forever — at least until your construction loan has started. Your only choice will probably be hard or private money, which can cost 5 percent of the loan upfront and 10 percent annually. This increase compared to institutional lot loan pricing is an expensive price to pay for money you already had in your pocket to begin with.

Cash reserves are required for construction loans

Banks want you to have cash on hand before they give you a loan. The amount of required reserves varies from bank to bank (see Chapter 9 for specific details). If you're short on the bank's cash requirements, it won't give you a loan — even with a paid-off lot.

Fending off the taxman

Real estate interest is one of the few deductions left for the average taxpayer. As long as you finish the house as a primary residence or second home, you should be able to deduct the interest and points paid on the loan. That means the government will pay a good portion of the payments for you to keep cash in your pocket. Discuss this option with your CPA or financial planner. You can ignore the tax benefits if you feel paying more than your share is part of your civic duty.

Chapter 4

Defining Your House Style

. . . .

. . . .

People tend to judge houses just like books — by their covers. Looks matter: People either fall in love at first sight or think a house is an eyesore. What individuals see is the home's style. That's why defining your home's look is so important, which is exactly what we help you do in this chapter.

You may have a truly unique home in mind — like one built of log or post-and-beam construction. This chapter offers more information on buying and building these specialized homes.

Getting to Know Your Style Preferences and Limitations

Close your eyes and picture the outside of your ideal home. Often, a definite look pops into mind: a yellow farmhouse with white shutters and a front porch, a boxy glass and concrete contemporary, or maybe a brick colonial with classic trim and an unadorned face. If an image of the perfect house doesn't immediately come to you, get off the sofa, find your camera, and get ready to go hunting.

Educating your eye

Get ready and get set to go on a house safari. It requires just a few hours, so you don't need to pack water or hire a guide. As you drive through neighborhoods, take pictures of homes that strike you — the good, the bad, and even the ugly. After you're back home with photos of a variety of house styles, you can consider what appeals to you and what doesn't.

If you can't find a variety of homes nearby to photograph or aren't impressed by the homes you come across, head to your local bookstore to stock up on a variety of magazines and books for inspiration. Look for

- General home magazines, such as *Architectural Digest, Better Homes & Gardens, House & Garden, Home, Country Home,* and *Country Living*

- Specialty home magazines, such as *Natural Home, American Bungalow, Old House Journal, Dwell,* log home magazines, timber frame home magazines, and Victorian-style magazines

- Regional magazines that cover your area, such as *Sunset, Southern Living, Down East, Midwest Living,* or *Coastal Living*

- Books that help you identify architectural styles, such as *What Style Is It? A Guide to American Architecture* by John C. Poppeliers (Wiley) or *A Field Guide to American Houses* by Virginia McAlester and Lee McAlester (Knopf)

As you flip through your new library of magazines and books, rip, clip, or photocopy photos that illustrate exterior details you like. Take note of what speaks to you: Is it a particular color or material? Window size, shape, or grouping? Shutters? Dormers or roof shape? Collect your clippings in a three-ring notebook or expanding file folder. Soon you can define your own personal style.

Discovering your local style

Now breathe deeply and take a reality check. Maybe you've fallen for an Asian-inspired pagoda-style home with a flat roof. It may not be your best choice, however, if you plan to build in an area with heavy snowfalls (that is, unless you enjoy the thought of several tons of wet snow crashing through your roof). This appeal to common sense will lead you to another rich source of home style ideas: your area's traditional (or *vernacular*) architecture. To discover this style, take a look around and note how older homes were built.

Regional styles typically evolve as a sensible response to a local climate, making them wonderful guides for what kind of buildings will work and what kind won't. Think about it: If that 200-year-old farmhouse down your street didn't know how to stand up to your region's weather, it wouldn't still be around today.

If you're moving to a new area to build your home, seek out the old-timers. In New England, for instance, the compact Cape Cod style home with its steep-pitched roof sheds snow and provides a snug second story under the roofline. These buildings offered warm shelter for early Americans living through harsh winters. Down south, in hotter climates, high ceilings in sprawling plantation homes lured warm air up and out of the way. Their wide porches gave folks a place to sit, eat, and often, sleep, outside.

Learning from years of building experience can help make your home easier to maintain and much more pleasant to live in. Employing vernacular architecture also provides the added advantage of making your home look like it belongs. Even if you want your custom home to be unique and make a bold-style statement, you don't want to own the neighborhood laughingstock either, and, at some point, you (or your heirs) will want to sell it. Then, your home's classic good looks will be a valuable asset.

Still, it's a free country, and if you live in Pennsylvania and really want a Southwest pueblo-style home with a flat roof and adobe walls (and you've ruled out a move to Arizona), you just have to be ready to pay more to build and maintain your home.

Playing by community rules

If aesthetics or the wisdom of your ancestors doesn't influence you in defining your home's style, your neighborhood might — in the form of covenants, conditions, and restrictions (also known as CC&Rs). These legalities (which you must agree to as a part of your real estate contract) can significantly impact your home, so do your homework before buying land. If you really want a log home, a barn full of llamas out back, or a three-story home overlooking a lake, you better read the CC&Rs' fine print before you buy your lot. Policies set out by homeowners' associations or local jurisdictions could restrict many aspects of your home, such as

- ✔ The height and size of your building
- ✔ Its proximity to a body of water
- ✔ The building materials you can use and your home's exterior colors
- ✔ Your ability to raise livestock
- ✔ The types of vehicles that can be parked on-site

Obviously, know upfront what you're in for. Some associations strictly enforce their restrictions. Others are more lenient. The upside of these kinds of restrictions is that they apply both to you and to your neighbors. You can rest assured your neighborhood will retain its current standards and be free of worry that some guy across the street will make his front yard a dumping ground for derelict plumbing fixtures. If you don't like these kinds of rules and regulations, you'd better buy your land in a different neighborhood.

If you plan to build in a community with CC&Rs, an architectural review board may have to approve your home's design. Because this review process may take months, add extra time into your construction schedule. (We talk more about review boards and the plan approval process in Chapter 6.)

Tapping the wisdom of the pros

So, you've looked at houses nearby, considered vernacular architecture, and flipped magazine pages until your fingers were raw, but still can't decide on a style. Think back to your house safari and remember the master: Marlin Perkins from TV's *Wild Kingdom*. When things got really dicey, wise old Marlin sent his sidekick, Jim, out to wrestle the wildebeests, while he stayed in the safety of the vehicle. Take a lesson from Marlin: Instead of risking a serious mistake, call for professional help. If you can't envision a style for your new home, or you and your spouse can't agree on a style, ask for input from an architect or designer. (See Chapter 5 for the specifics of finding and working with an architect.)

Your design pro will look at the scrapbook you have compiled and ask questions about your lifestyle. She may suggest a home style that's common to your area or make the case for something more unusual. During this design process, you can help by

- ✔ **Openly discussing your budget.** It wastes your time and the designer's if you hedge about how much you want to spend on your home.

- ✔ **Offering honest feedback.** If your designer seems to be veering off track, say so. The goal for everyone involved should be to create the best possible home.

- ✔ **Keeping an open mind.** Don't dismiss an architect's suggestions without giving them some consideration.

At some point in the process you'll wonder if hiring a designer or architect is necessary or worthwhile. The fees these professionals charge typically fall in the range of 10 to 15 percent of your project's total cost. If you can find a plan in a book that suits your needs as well as your lot and your neighborhood restrictions perfectly, you don't need an independent designer.

But remember those wildebeests: By not using a designer or architect, you're facing some wild times all by yourself. An architect or designer often can act as your advocate in a confrontation with a contractor. Because the designer's job is to be knowledgeable about building materials, he can save you from spending money on a product or material that simply won't work in your area or in your home. Furthermore, he can create a home that truly fits on the land you worked so hard to find.

Good communication and rapport are essential elements in your relationship with a design professional. If these two elements evaporate over the course of your project, you may need to cut ties with your building or design professional and choose someone new. This action is drastic, however, and should only be taken if you have honestly tried and simply can't reconcile your differences.

Considering Conventional Construction: Wood versus Steel

By far, the most popular home construction technique is *conventional construction,* which uses vertical studs to create the home's skeletal system of both exterior and interior walls. Choosing conventional construction allows for a wide range of styles; the studs are simply the basic ingredient. The studs in your walls will be either *dimensional wood lumber* (lumber that has been cut to specific, standard sizes) or steel.

More than likely your house will be stick-built: It will use good old-fashioned wooden *studs* — long, thin boards used throughout the framing process. Wooden studs are popular because

- Most framers have the necessary tools (hammer, nail gun, saw, and so on) to work with wood.
- Wood is mass-produced and costs up to 30 percent less than steel.
- Most subcontractors and laborers know how to work with wood.
- You can alter wood-framed walls relatively easily in the future.

A second option — steel framing — has been in use for office buildings for some time now. For a variety of reasons, steel is finding its way into more residential construction projects. Although steel isn't for everybody, it does have a few advantages over wood:

- ✔ Steel offers the greatest strength for the lowest price of any building material.

- ✔ Steel is inorganic. Galvanized steel doesn't burn, warp, rot, split, crack, creep, or get eaten by termites and other creepy crawlers.

- ✔ Steel is dimensionally stable. It doesn't expand or contract due to moisture.

- ✔ With steel, you have less scrap and waste (2 percent for steel versus 20 percent for lumber).

For every person singing the praises of steel framing, another is swearing that wood is the only way to go. The benefits and disadvantages seem about equal for each method, so it comes down to what makes the most sense to you and your contractor, the price, the climate, and other considerations. After you find a contractor you like and who you can afford, ask for references and check out his work firsthand. Then we advise that you yield to your builder's preferences on the wood versus steel issue.

Enjoying the Warmth of a Log Home

Everyone who has ever watched *Bonanza* or read a *Little House on the Prairie* book (come on, admit it!) has harbored a fantasy of living in a log home. For some, the fantasy is fleeting; for others it becomes a lifelong obsession.

Building a log home does bring a variety of advantages:

- ✔ You have a range of log home styles available to you. Logs can do justice to country, rustic, Victorian, Arts & Crafts, and even contemporary styles.

- ✔ You'll end up with a custom home that is distinct and filled with character.

- ✔ The thickness (also known as *thermal mass*) of logs offers good insulation and helps the home retain warmth in the winter and remain cool in the summer.

- ✔ Many log home owners say their homes are quieter inside than conventional homes.

- ✔ The rugged good looks of a log home make it fit naturally into a variety of settings, from prairie to woods and from lakeside to mountaintop.

But if you chose to build a log home, understand that you're asking for something out of the ordinary, which may cost you more than a conventionally built home in terms of time and money.

The following sections briefly outline how the log home–building process differs from the conventional building process and also provides advice on selecting a log producer if you decide to make your log home dream a reality. If you're interested in a log home, we suggest that you continue your search for more information in several places:

- ✔ **Use your friendly, local bookstore or library.** Look for books and magazines on the subject. On the newsstand, look for *Log Home Living, Log Homes Illustrated, Country's Best Log Homes,* and *Log and Timber Style.*

- ✔ **Stop by log model homes in your area.** Talk with the sales representatives and ask questions.

- ✔ **Knock on the doors of log homes you see.** Log home owners are typically quite happy to discuss their home–building experiences.

- ✔ **Talk to real estate agents, your lender, insurance agents, and local builders about your desire for a log home.** These building professionals may be able to point you in the right direction.

- ✔ **Finally, keep an eye out for the dozens of log home shows that are held around the country every year.** A show is a good place to see firsthand what various log home producers have to offer. (You can find information at `www.loghomeliving.com` or `www.loghomeexpo.com`.)

Two ways to skin a log

As you research log home options, you'll find two basic types of log producers. Most companies are *manufacturers* that mill their logs using machinery. The end result is logs that are uniform in shape and dimension. Manufacturers typically also use machinery to cut the corner notches that connect the logs. Many manufacturers offer log siding that can be used on a home's exterior (and also on interior walls) to give the look of log construction with a slightly lower price tag. The remaining log companies call themselves *handcrafters.* As the name implies, these companies employ crew members who use chain saws, or sometimes even hand tools, to shape the logs for homes one at a time.

Which should you choose? As with so many decisions involving custom homes, the answer is, "It depends." Consider the following:

- ✔ **Do you need your home fast?** Some handcrafters are tiny operations that produce only a few homes a year. If you want to build quickly, choose a larger production handcrafter or a manufacturer. Remember, though, that even these producers need weeks or months of lead time to cut your logs.

- ✔ **Do you prefer a certain style of log?** For some people, only square logs accented with wide bands of white *chinking* (the material used to seal the joints between logs) look like a "true" log home. Other people prefer smaller, rounded logs with no chinking. You can see an example of milled logs in Figure 4-1. The look you like can help you decide on a producer.

- ✔ **How much log is enough?** Do you want massive, 20-inch diameter logs for your home and the interior partition walls, too? Most manufacturers' machinery simply can't handle oversize logs, so if you think bigger is better, you'll most likely choose a handcrafter.

- ✔ **Will it play in your neighborhood?** If you plan to build in an area filled with log homes, consider following local traditions. Are most of the homes built with huge, handcrafted logs? Choosing smaller, machined logs may make your home the odd man out.

- ✔ **Can you pay for it?** Does a handcrafted home cost more than a manufactured home? Again, it depends on whom you ask. Handcrafted homes are preassembled in the crafters' log yard, so reerection of the shell on the construction site may go faster, saving time and money. Still, handcrafting a log home is labor intensive; after all, you're asking someone to build your home *by hand.* Manufactured logs may take longer to assemble on-site, but producing them is more mechanized (and less expensive). The only way to find out is to ask plenty of questions and compare ballpark estimates on your home from both types of producers.

Figure 4-1:
This log home is made from milled logs interlocked together.

Courtesy of The Original Lincoln Logs LTD.

Thinking that logs are a money-saving, do-it-yourself alternative to a conventionally built home? Do your research. Even if you mastered your Lincoln Log kit at an early age and know which end of a chain saw to point toward a tree, buying logs from a supplier and planning to cut, notch, and stack them yourself may provide short-term savings at the cost of long-term headaches. Unless you have a sound plan (or know a builder who has good experience) to stack, fasten, and seal the logs into walls that will pass building codes and stand up to the elements year after year, buy a log package from a reputable log producer.

Purchasing your log package

Most log manufacturers and some handcrafters sell log packages. A log package assembled by a log producer will most likely come with fasteners to hold the logs together, precut notches to interlock the logs at the corners, some kind of insulating material to place between the logs, sealants for weatherproofing, and instructions for your contractor (or you) to put it all together.

You can purchase most log packages through a network of dealers or local sales representatives. Your local representative can help you purchase your log package and, if she is a general contractor, may also help you build your home. As your liaison with the log producer, your dealer can help you with your home's design, coordinate the logs' delivery, and walk you through the purchasing and construction processes.

One major difference in the construction process occurs in the loan process. Understandably, log producers don't want to cut and process logs for your job without knowing that you're serious about building the home. For that reason, they ask for a large deposit (typically 50 percent of the package cost) before they cut the logs. The construction lender you choose must be willing to advance that sum from your total construction loan early in the process. (For more information on the construction lending process, see Chapter 10.)

After the log producer delivers the logs to your building site, the producer's crew or your general contractor's crew will stack them into walls. The home will take shape quickly, because after they're stacked, the logs create finished walls, both outside and inside. They don't need to be insulated, covered with drywall, or sided on the outside. Log homes do have some unique construction steps, but generally subcontractors working on a log home will proceed as they do on a home built with conventional construction.

Highlighting Wood Inside — Timber Frame or Post-and-Beam

If you love the look of wood but not necessarily log homes, you may opt for a post-and-beam or timber frame home. Chances are you've seen this kind of building method used in old barns or churches. A frame of substantial, interlocking timbers supports timber homes. The timbers may be held together with pegged joinery cut out of the timbers themselves or fastened with bolts or metal plates. Unlike log homes, from the outside, timber frame or post-and-beam homes can look like any other home. Inside, the frame's beauty is revealed, adding character and wood tones to interior spaces. (See Figure 4-2.)

Timber framing is a new take on an old tradition that has several benefits:

✔ Timber framing creates a home with high energy efficiency. The energy efficiency comes with the method of enclosing the frame. Many timber frame companies suggest the use of structural insulated panels (SIPs) to wrap the frame in a continuous envelope of insulation.

Figure 4-2:
The beauty of wood timbers is revealed on the home's interior where posts and beams allow for open, vaulted spaces.

Courtesy of The Original Lincoln Logs LTD.

SIP: Building a better sandwich

Structural insulated panels (SIPs), which range in size from 4-x-8-foot rectangles to large 8-x-24-foot sheets, are "sandwiches" of expanded foam between sheets of plywood, oriented strand board (OSB), or other solid material such as drywall. SIP manufacturers bond the layers together using pressure and industrial adhesives. After they're precision-formed in a factory, the panels are ready to be shipped as is, or the manufacturers' crew can cut the panels to accommodate windows and doors. The panels' plywood or OSB surfaces serve as the base for drywall or paneling in the home's interior and

any common type of finish on the exterior, such as siding, brick, or wood shingles. During construction, electricians can run their wires through chases drilled in the panels' foam or simply install wiring behind wall baseboards. To connect the SIPs to the frame, the installation crew spikes or fastens the panels to the timber frame, and then fastens them to each other. SIPs can be used for roofs and floors as well. SIP walls may cost a bit more than conventionally framed walls, but their increased energy efficiency should make up for any additional upfront cost.

✔ Timber framing, which has been practiced for centuries, uses large wood posts and beams that add plenty of character to a home.

✔ Through the use of trusses, timber framing can create homes with large open spaces that span great distances and accommodate the open floor plan that many homeowners desire.

As with log home construction, timber framing may add to a home's bottom line. It's a specialty home, after all. You may be able to recoup the extra money in energy savings over the life of the home and in a higher price set at resale.

To find out more about timber framing:

✔ **Go to your bookstore or library.** Look for books and magazines on the topic. Look for *Timber Homes Illustrated, Timber Frame Homes,* and *Log and Timber Style* magazines. Tedd Benson's books, *Timberframe: The Art and Craft of the Post-and-Beam Home* and *Timber-Frame Home: Design, Construction, Finishing* (both by Taunton Press) are good starting points.

✔ **Search online for timber frame and post-and-beam companies to see photos and descriptions of their work.** Many companies sponsor open houses, workshops, or frame raisings that are open to the public.

✔ **Visit the companies' model homes or homes of their previous clients to see framing in person.** Ask the representatives or homeowners questions.

✔ **Attend one of the dozens of log and timber frame home shows held each year around the country.** You can find information at `www.loghomeliving.com`, `www.timberframehomes.com`, `www.timberhomesillustrated`, or `www.loghomeexpo.com`.

As you research timber framing options, you'll find that several different kinds of companies offer timber frames and post-and-beam packages. Most packages contain the timbers for the frame, and oftentimes, the system to enclose the frame.

The style of the timbers can create or enhance a number of different styles, from Shaker style to Old World lodge to Arts & Crafts style to something more rustic. If you're an antiques lover, you might search out a timber producer that can supply timbers salvaged from old buildings or a complete frame from a former barn to give your new home a vintage feel.

If you're concerned that timber framing will bust your budget, opt for a *hybrid* home. In a hybrid, timber framing is used selectively, typically in large public areas, like great rooms, entryways, or dining rooms. Smaller rooms, utility spaces, and bedrooms can be built without timber framing to save money.

Along with cutting the timbers, and often raising them on-site, companies may offer additional services. Make sure you understand if your timber producer provides these kinds of services, or if you need to hire additional, local help to work on your project. The companies may

✔ Enclose the frame with SIPs, conventionally framed walls, or some other enclosure system

✔ Have a staff member who can design the frame or the entire home

✔ Serve as the general contractor for your home's construction from start to finish

In your research, you'll find that timber frame operations range in size. Some small companies concentrate on handcrafting just one or two frames a year. Other timber frame and post-and-beam companies produce many more homes annually and sell packages through sales representatives or dealers. Some of these representatives live or work in a model home. Your representative can help you purchase your timber frame or post-and-beam package, and, if he or she is a general contractor, may build your home. As your liaison with the company, your dealer can help you with your home's design, coordinate the frame's delivery, and explain the purchasing and construction processes.

One major difference in the timber home construction process occurs in the lending process. As with producers of log homes, a timber producer doesn't cut a frame without knowing you're serious about paying for it. For that reason, the producer will ask for a large deposit (typically 50 percent of the package cost) before the timbers are cut. The construction lender you choose must advance that sum from your total construction loan early in the process. (For more information on construction lending, see Chapter 10.)

After delivering your timbers to your site, your timber producer's crew will assemble and raise the frame. The framer's crew, or your general contractor's crew, will then enclose the home and it will be considered "under roof" or "dried in." Although timber frame and post-and-beam houses have some unique construction steps, subcontractors working on these homes generally proceed as they do on a home built with conventional construction. As with other types of custom home construction, the general contractor typically supervises the entire timber home project from start to finish.

Considering a System Approach

When you need new kitchen cabinetry, you usually don't ask a cabinetmaker to haul his tools, crew, and raw materials to your home to build cabinets for you on-site. You pick a style at a cabinet supplier's showroom and order the number and size you need. The supplier sends your order to the manufacturer, who goes about building your cabinets in a factory. You can apply these very same methods of production and delivery to building a home. Homes produced in a factory are called *system-built homes*.

Before visions of double-wides rusting in trailer parks come dancing into your head, you need to know that "trailer homes" are now labeled as "manufactured homes" and are built to a different code (known as "HUD code" for the Housing and Urban Development agency that writes it) than system-built homes (which are built to the same codes as site-built homes). System-built homes include both modular and panelized homes.

Weighing your options

Today's factory-built or system-built homes come in many forms, but fall into two broad categories:

- **Modular home:** Built using preconstructed sections of the house, called *modules*
- **Panelized home:** Built using preconstructed wall, floor, or roofing units, as well as all other components of the house

Each form uses components that a supplier cuts to fit — and sometimes even assembles — and then delivers to the building site for completion.

Some companies specialize in creating large "chunks" of homes in their factories. These companies build panels or wall sections using conventional construction or SIPs that are made of layers of plywood or oriented strand board and foam. The panels may contain prehung doors or fully installed windows, and they're shipped to the construction site to be joined together to create walls on the foundation.

Why would you choose to have your home prebuilt instead of built on your property? The question for many is, "Why wouldn't you?" Some of the benefits include the following:

- System-built home plans are as flexible as conventional construction homes in that they can have one, two, three, or more levels, gourmet kitchens, finished basements, brick walls, and fireplaces. And, no, they don't have wheels.

- Prebuilding in a factory saves time because workers aren't delayed by weather or lack of available materials.

- The volume of building at the factory means workers can be concentrated in one place — painters can stay busy in a modular home factory shift after shift. They don't have to travel from job site to job site or wait between jobs.

- Inspectors in the factory ensure that work is done precisely to plan and with closer tolerances than on a job site.

- Best of all, you'll likely save money by choosing a modular or panelized home. Here's where the savings come in:

    • There is less risk that valuable building materials will be stolen or damaged on the construction site.

    • A shorter construction time means fewer interest payments on your construction loan.

    • Design work or drawings may be included in the price of the home package so you can avoid paying an independent designer or architect.

    • Because you can't make changes to the structure after it reaches your site, you don't rack up expensive change-order fees.

Still, a modular or panelized home may not work for you for various reasons. Access to your site might be impossible for large delivery trucks. Your home's designer, architect, or contractor may prefer to work with conventional site-built methods. The level of amenity you want for your home may just be too high to be efficiently fulfilled by a system manufacturer.

If you're considering a modular or panelized home:

- ✔ **Start with the companies' floor plans.** Review them and their list of construction specifications.

- ✔ **Make sure your site is accessible to the trucks that will deliver your home's components.** Speak with your representative to make sure your site is appropriate.

- ✔ **Begin looking for a general contractor to oversee the building of your home's foundation, the installation of the prefabricated components, and the final finish work.** If you want to serve as your own general contractor, ask the manufacturers you're considering if they have worked successfully with owner-builders in the past.

- ✔ **Tour as many system-built homes as possible.** You can see firsthand how the finished structures look and feel.

- ✔ **Inspect each company's finished product closely.** Ask questions of people who have lived in the houses for a few years or longer. Have the homes stood the test of time? Are the homeowners experiencing any problems with their homes?

- ✔ **Check out appropriate Web sites.** For fast facts about system-built homes, compiled by the Building Systems Council of the National Association of Homebuilders, go to `www.nahb.org/generic.aspx? sectionID=455&genericContentID=10216`.

Making a purchase

If you have your home system-built, you'll first contact either a local builder who has experience putting the houses together or a local system-built home manufacturer representative. The representative or dealer you purchase your package from will have specific assembly and construction instructions for your contractor. Many manufacturers provide training courses for complex systems, and in some cases, the representative also serves as a contractor.

You need to find a company that can provide you with two things: A home design that fits your needs, or that can be altered to fit your needs, and a home built to the level of amenities that you desire.

Not sure what design works best for you? Our advice is to study each manufacturer's stock designs and then ask if the manufacturer has the ability to customize the plans. To determine the quality of a manufacturer's homes, ask to see the company's specifications. Some areas to take note of include

- ✔ **Insulation:** The manufacturer builds your home to meet the insulation level required by your local building codes. If you want greater insulation, talk to your manufacturer, and expect to pay more.

- ✔ **Pitch of the roof:** A flat pitch makes the home look more like a trailer-park model and less like a custom home. Steeper pitches look richer.

- ✔ **Roofing and siding materials:** If a manufacturer can't provide a certain type of shingle or exterior finish for your walls, you can have your builder apply the shingles after the home is on-site.

- ✔ **Stairs:** The standard stairs may be carpet-grade. In other words, you need to cover the stairs with carpet. If you want exposed wood on your stairs, ask what options are available. Again, be prepared to pay for an upgrade.

- ✔ **Types of appliances:** Can the manufacturer accommodate your lifelong desire for a commercial-size range in the kitchen and a high-end, built-in refrigerator? Some can, but be prepared to pay extra.

- ✔ **Windows:** Ask about your options for windows, and be sure to see the window units firsthand. Be sure the windows are easy to open, close, lock, and tilt for cleaning (if that's a feature you prefer). For more in-depth information on comparing window quality and energy-efficiency, visit the EnergyStar program's Web site at `www.energystar.gov/index.cfm?c=windows_doors.pr_windows`.

If you have firm ideas for your new home, make sure the manufacturer you choose can supply what you need. For example, do you have your heart set on a vaulted master bedroom? Because they are limited by what size components can be delivered by truck, not every modular manufacturer can build a home with ceilings above 8 or 10 feet. Ask questions and be sure to look at homes the company has previously built.

Perhaps you want to bring your own drawings to the table. Some modular manufacturers can accommodate custom-drawn plans, but you may find that panelized manufacturers will more readily accept plans from an architect or independent designer.

On the line

With modular homes, workers install wiring, plumbing, ductwork, cabinetry, and some plumbing fixtures in the factory, and also apply certain finishes, such as paint, flooring, and countertops. After the workers finish, they shrink-wrap the modules like giant pork-chop packages and load them onto flatbed trucks for delivery.

Before you hit the road, Jack, be sure you know the rules. Some states limit the weight of items that can be shipped by truck at certain times of the year. Imagine that a truck carrying your kitchen jackknifed on an overpass in a blizzard, or worse, overturned. Discuss your construction schedule and any possible restrictions with your contractor or sales representative well in advance of your purchase.

After the house arrives at your lot, workers cut away the heavy-duty wrapping from the modules, and a crane lifts the modules off the trucks and on to the home's foundation. Depending on its size, the home may comprise two, three, or more modules. Under the supervision of your builder or manufacturer's representative, the crew completes the steps necessary to make the home ready for occupancy.

A modular home leaves the factory pretty close to complete. After the pieces arrive on the construction site, the homeowners can usually move in within just a few weeks. Make sure you're packed and ready to move in.

Panelized homes go up in much the same way, except that the parts that are delivered to the site are smaller and require more assembly on-site. A crane may still be needed for lifting large wall or roof sections. After the pieces are in place, the home is ready for interior finish work. The construction time may vary from one system to another. Obviously, those systems that leave the factory in more-finished states require less time and labor on-site.

Unearthing Alternative Construction Methods

Does a system-built home sound too buttoned-up for you? If so, then get earthy with alternative building methods. Some of these methods are ages old and recently rediscovered; others have become viable and more widely available by recent technological improvements. Two different options include

- ✔ **Straw bale homes:** Today, people are rediscovering the ability of straw bale construction to create highly energy-efficient homes. After the bales are stacked into walls, they are coated outside with an earth-based material and inside with plaster, and then finished as any other home.

  Straw bale homes usually resemble adobe homes with thick walls and gently rounded arches. They offer their owners great efficiency and a quiet interior. An added bonus is knowing that you created something lasting out of a material that would normally be burned or sent to the landfill.

  In laboratory tests, the bales have resisted damage from fire, mainly because of their density. Insects don't seem to be a problem and moisture issues aren't of great concern as long as the bales are well dried before construction and kept dry until they are finished on the exterior. Check with your local municipality and lenders if a straw bale home sounds like your cup of tea.

✔ **Rammed earth homes:** Like straw bale homes, *rammed* (or stabilized) earth homes use a natural material to create a cozy home with thick walls. In this type of construction, your wall contractor mixes *screened* (sifted) soil with cement and water, and then pours it into wall forms built on your site. The crew then uses pneumatic tampers to compress the earth mixture in the form. After the mixture sets, the forms are removed and the 18- to 24-inch walls are complete. The walls can be left as is, colored with pigments, or sealed with stucco.

As with other building forms that create thick, solid walls, rammed earth homes enjoy increased energy efficiency and quiet interiors. Solid, natural material walls paired with a heating system that doesn't require blowing air provides an ideal home for those with heightened sensitivity to chemicals or synthetics.

Keep in mind that although the raw materials for a rammed earth home — notably dirt — are widely available and, well, dirt cheap, the labor involved adds to the home's cost. Knowledge is the best weapon when encountering setbacks, so visit rammed earth suppliers online or in person and talk about your plans with homeowners and builders who have relevant experience.

If you're interested in an alternative building method or material for your custom home

✔ **Do as much research as possible.** Look for magazines like *Natural Home* and books such as *The Art of Natural Building* by Joseph F. Kennedy, Michael Smith, and Catherine Wanek (New Society), and *Alternative Construction: Contemporary Natural Building Methods* by Lynne Elizabeth and Cassandra Adams (Wiley).

✔ **Get help.** Enlist the aid of experienced builders or suppliers.

✔ **Attend workshops and seminars.** Many producers or building schools offer workshops and seminars, which are particularly valuable if your home will be a do-it-yourself project. Knowing in advance if you have the skills and persistence to tackle the job is better than finding out too late.

✔ **Look online.** Check out `www.greenbuilder.com`, `www.thelaststraw.org`, or `http://oikos.com/index.lasso` for further information on an array of alternative building methods.

Believe it or not, you *can* find financing for a home built with alternative building materials. It may, however, take more legwork than finding a lender to finance your purchase of a regular suburban tract home. Start by talking with the company that's helping you build the home. Most likely, the company's previous clients have faced the same situation, and they may be able to offer sound advice. You can also search online for lenders who frequently make construction loans. Having complete plans for your home, information on the building system from an experienced builder or home producer, and a clear idea of the costs of your project can help ease a lender's mind and make your project that much more appealing to a loan underwriter.

Living off the grid

For some, the idea of living without public utilities or supplementing those utilities with wind, water, or solar power sounds like roughing it. But technological advances in these alternative power systems make it easier for people who wouldn't consider themselves pioneers to make their own energy.

The use of alternative energy sources probably won't affect your home's style except in the instance of passive solar heating. Homes designed to benefit from solar energy in this way are carefully sited on their lots. Banks of windows bring sunlight into the home and focus the warmth into the home's *heat sink* — typically a masonry wall or floor that collects the warmth, and then slowly radiates that warmth back into the home. Of course, if you use active solar systems, such as panels of solar cells for generating electricity or large solar collectors for heating water, you'll need to either hide them or integrate the look into your home design.

If you're interested in using alternative energy to power your home, research your choices thoroughly. Start with books and magazines, and then supplement your reading with visits to the Web sites of suppliers of solar, wind, and water energy equipment.

Although building a home powered solely by alternative energy (called *off-grid*) is good for the planet, it may cripple your checkbook: Homes that are built off-grid aren't eligible for bank financing. If you need a loan, connect your home to the grid, and then simply use your alternative energy system. Many people who make their own power, but who are still connected to the public utility grid, are actually able to "sell" their excess power back to the utility company. Rules and regulations on this buy-back of power vary from region to region. For more information, visit www.oikos.com.

Chapter 5

Architects and Design: Time Spent Is Money Saved

In This Chapter

▶ Using an architect

▶ Relying on other design resources

▶ Placing the house

▶ Making aesthetic choices

▶ Defining the details

Designing your own home is probably the reason you started on this custom home journey in the first place. But, guess what? Designing your own home also is the most complicated and challenging part of the process (although at the same time it can be the most rewarding). Wondering what the biggest mistake made in the design process is? Underestimating the time you need and the sheer amount of decisions you need to make.

In this chapter, we assist you with the multitude of decisions and choices that you'll face by giving our insights from our own years of experience. We start with the discussion of using an architect, relying on a stock plan, or designing your new home yourself. We give you an introduction to choosing styles and determining functionality. Finally, this chapter has lists of features and choices to use when making the necessary decisions for your custom home. The result is a script for you to follow in conversations with your architect and contractor.

You're designing *your* custom home project. There is no right or wrong way. But some ways may be easier or less costly than others. Entering into the design process with open ears and an open mind is better than resisting new ideas. Be prepared to take extensive notes during the process because it occurs over an extended period of time, and you may need to refer back to what you were talking about six months ago.

Arming Yourself with an Architect

Like any other professional consultant, because you're paying the money, you have the right to determine how much your architect will be involved in your home-building project. Architects can serve you in several ways. They can take a stock plan you have seen in a magazine or on the Internet and simply modify it to fit your needs, or they can help extract ideas from your mind and create a whole new home to meet your dreams. Some people simply want an architect to design a home based upon their thoughts and needs. Others want to be fully engaged in the entire experience and use the architect as an interpretive tool, expressing what they see completed in their own mind.

Deciding whether you even need an architect

The question of whether or not you need an architect for your project boils down to two elements: time and experience. If you're in no hurry to move your project forward and are willing to invest the time to figure out all the ins and outs of the process and make the right choices, then an architect may be unnecessary. But if you work for a living, are raising a family, or don't have the slightest inclination to take the time to figure out design, construction, and building codes, then an architect will be a welcome addition to your custom home team of professionals. Here is a list of questions you need to ask yourself to determine if you're up to the task of designing your own home:

✔ Does your state require an architect for submitted plans?

✔ Will your project require extensive structural engineering?

✔ Are you extremely picky and difficult when making decisions?

✔ Are you lacking in aesthetic vision?

✔ Do you have difficulty understanding home functionality?

If you answer yes to any of these questions, then you'll probably gain value from an architect. Despite their seemingly high cost, architects can save you time and money by bringing their experience to the table. Their insights on functionality and government bureaucracy can save you months of time and thousands of dollars.

The biggest question to ask yourself is whether you have the confidence to take this project to its completion. You may prefer to dabble or play with the design aspects, but an architect is a true professional who has spent years becoming an expert at home design. If you were to go alone without an architect, you might spend a great deal of time and energy gathering the information that already sits at the architect's fingertips.

Finding the right architect

Locating an architect is as easy as picking up the phone book or searching on the Internet. The hard part is figuring out which architect is right for you. You have several approaches you can take to find the right architect:

- **You can drive through neighborhoods of custom homes looking for houses that strike your fancy.** Don't be afraid to knock on the door and ask for the architect's phone number. Most people are happy to share the information while you're flattering their home.

- **If you're lucky enough to have many friends with custom homes, you can ask for referrals from them.** What are friends for?

- **You can hit the Web.** The American Institute of Architects (AIA) has a list of its members by location at www.aia.org. Click on the "Architect Finder" option, enter your zip code, and select the "A Home for Yourself" building type to find architects in your area.

Now that you have a list of prospects, you need to compare them. Cost is usually the first comparison but by no means the most important. When choosing the right architect, look for someone that fits the needs of your particular project and working style. You may want someone who manages the whole process or perhaps will work with you in a teamlike manner. Plan to have several discussions with two or three different architects so you can choose the right one for you. Here are the important issues to address in those discussions:

- **Aesthetics:** You need to see if the architect can create something that suits your taste. Ask to see many of her prior designs. Ask for introductions to the owners. Go to the completed houses and see if the floor plans make sense and are comfortable for you. If you don't like the homes she previously designed, chances are you won't like a new one either. A good architect is also a good listener. Look for someone whose taste is similar to yours and who will design what you're looking for. You want your new home to be a reflection of you — not a monument to the architect.

- **Experience:** What's the point in hiring an expert who knows less than you do? You want an architect that has designed many custom homes and is familiar with the process. A commercial architect who specializes in office buildings may be looking for the next new challenge, but his lack of residential experience could create problems for you down the line with builders and planning departments. An architect needs to have a minimum of 15 custom homes under her belt to be considered for your project.

> ✔ **Local knowledge:** Every municipality and planning department is different in the way they process custom home plans. Some are more bureaucratic than others. Much of the architect's time may be spent working your plans through the system. If you're looking for exceptions (*variances*) from the established local guidelines, you could have a fight on your hands. This fight could cost you time and money. An architect with local knowledge and experience can save you from costly battles and exercises in futility. (Check out Chapter 6 for more information on local design guidelines.)

Managing the architecture process

If you're lucky, you may find the perfect architect — someone who is attentive to your needs and makes the process easy. In a perfect world, the architect would come up with the perfect design first time out with a minimum of communication. Sadly, wake up and smell the coffee: You don't live in a perfect world. Most custom home architects are small businesses; they're shoestring operations without huge profit margins. They tend to be overloaded with work and less concerned for your time frames than you may be. The more successful the architectural firm, the busier it will be. You need to manage your expectations and the process.

Set your initial meeting as a getting-to-know-you session to get a feeling for how you'll work together and to see if you have a common vision for the project. After everything gets going, stay proactive in driving the process. (To stay on top of everything, you need to call the architect regularly to check progress and set the next appointment.) After all, it is your house and your time frame. The architect's job is to present you with information and decisions to be made. All the decisions relate to four basic elements:

✔ Aesthetics

✔ Cost

✔ Quality

✔ Time

Get the facts from the architect and conduct your own research. The more you take responsibility for educating yourself and making some decisions upfront, the greater chance you have of eliminating problems, saving some money, and getting the home you're looking for. Use your architect as the high-paid consultant that he is and make sure to set a regular meeting schedule with him. Doing so helps you get a better handle on the time and dollars involved to design the project. Setting a regular schedule also reduces the panic or inquiry calls that can cost you more money and frustrate you during the process.

What does all this cost?

The architect has the worst position of all the partners you have in the custom home process — the first position. Most people expect to put out a substantial chunk of money for the building lot, but the five-figure check you'll write to the architect is probably the first one you signed that didn't get you a big hunk of steel with tires and headlights. Don't let the price tag scare you; during this stage isn't the time to skimp. If you've done a good job of budgeting (see Chapter 2), you'll easily make up the initial costs later in the build.

Architects generally charge in one of three ways (or a combination of all three for various stages of the project):

 ✔ On an overall percentage of the build

 ✔ On an hourly basis

 ✔ On a fixed-price basis

The following sections provide some vital information you need to know about each billing program and which one is better for you.

Percentage of the build

The first way of billing is on an overall percentage of the build. This method can range anywhere from 3 to 10 percent of the total project costs, not including land. So a typical $500,000 custom project may cost you $15,000 to $50,000 in architect fees for the plans and the architect's time in getting the plans approved. In some high-end projects, architects may charge as much as 20 percent, equaling hundreds of thousands of dollars. This method doesn't have a set percentage, so you need to evaluate the value of the architect's services that you receive in exchange for your hard-earned money.

Hourly basis

The second way of billing is on an hourly basis. Hourly rates vary widely depending on the firm you engage, its experience and reputation, and its location. (A firm in Los Angeles is probably going to cost significantly more than one in Des Moines.) Expect to pay anywhere from $50 to $350 per hour, depending on these factors. You'll pay the architect for the following list of items to get you to the permitting stage:

 ✔ Construction documents

 ✔ Landscape plan

 ✔ Mechanical and electrical drawings (see Chapter 6)

 ✔ Plan copies ($4 per page)

- ✔ Soils report
- ✔ Structural engineering (see Chapter 6)
- ✔ Surveyor
- ✔ Time for the architect and associates

You can pay all these items directly through the architect, or you can pay for them separately on your own. Project costs can be more clearly broken down into time and materials (expenses), and they'll vary depending on where you live and on your project's scope. The consultants and types of reports they can generate vary due to the scope of work and the requirements of governing bodies, such as planning and building departments.

When paying by the hour, time is money. Use the architect's time for providing information and education. Keep all discussions or disagreements between spouses or partners at home, not in the architect's office while the clock is ticking. Absorb as much information as possible and take detailed notes to review on your own time. The more you prepare for the meetings with the architect, the more efficient those meetings will be. Shorter meetings mean less billable hours and less money out of your pocket.

Considering the design-build combo

A number of companies today design the homes and build them for you. Such companies are called design-build firms in the industry. Many custom home companies have added architects to their staffs so they can provide you with a seamless process from start to finish. The main advantage of this approach is consistent communication throughout the process. The builder has experience building what the architect designs and the architect designs a home based on the builder's expertise. You can take advantage of the cost savings attached to using one firm for both the design and the build, but you need to compare the price and work with that of independent architects and contractors before making a decision. You can find an annual list of leading design-build firms online at www.designbuildbusiness.com.

Just because they handle both design and build aspects through one firm doesn't mean you can reduce your investigation for finding the right partners for your project. Design-build can be something of a conflict of interest because it doesn't involve any competitive bidding in the process. Not only that, but the project's ultimate quality is tied up in the firm's ability to both design and build in a cost-effective standard. Pick a design-build firm using the same criteria we suggest for picking an architect in "Finding the right architect" section earlier in this chapter, and investigate the builder portion of the firm using our suggestions in Chapter 2.

Fixed-price basis

The third way of billing is on a fixed-price basis, where your architect quotes you a firm, all-inclusive price for the entire job. On one hand, this option can be beneficial to you because you know exactly what your architect will cost, but an architect may underestimate and you may suffer when he feels he has already put in too much time.

Always discuss the price with the architect before you sign a contract. Get a complete estimate upfront with a detailed breakdown of expenses. Set a maximum price with progress payments based upon certain milestones such as the preliminary design and design review approval (discussed in Chapter 6). Keep the communication open along the way so you don't encounter any surprises.

Looking at Architect Alternatives

Not everyone uses an architect when building a custom home for several reasons. For example, aside from money being a factor, you may have design skills yourself that you wish to exercise. No problem! Technology has improved the choices for designing a custom home. You can also utilize alternative consultants if you want to save on architecture fees. These resources require additional responsibility on your part and may still have additional costs depending upon the design requirements of your local government. The following sections explore architect alternatives.

Published floor plans — Picking a home from books or online

You can purchase thousands of plans from magazines and online resources — many of them quite good. The choice is endless. The magazine shelves at bookstores are stocked with more than ten new magazines every month; these plan books have houses to fit every size and budget. You can buy the preliminary floor plan and elevations for a few hundred dollars. You can also purchase complete building plans including the structural drawings from these sources for a few hundred to a few thousand dollars. (Check out Chapter 6 for a complete explanation of the differences in types of plans.) Even if you don't buy the full set of plans, the magazines and online sites make for good conversation starters with your family and architect. Here are a few of our favorite Internet resources for plans:

- www.familyhomeplans.com
- www.eplans.com
- www.dreamhomesource.com

Although buying plans may seem like a less expensive approach than using an architect, it depends heavily on your situation. Many of these plans don't include the foundation or full structural drawings required for permitting. The plans need to meet all the design and code regulations for your area, and if they don't, you'll have to hire an architect or engineer to make any changes necessary for permitting. Depending on those costs, you may or may not save money by using the stock set of plans compared to hiring an architect to design your home. Buying plans from a book can, however, be a great option if you're building on a flat lot with liberal design guidelines. Otherwise you may be wasting your money.

Software programs — Designing your own plans

A number of software programs exist for individuals wishing to design their own home on a computer. Many of these software tools make it easy with templates for rooms and architecture choices. For less than a few hundred dollars, these programs can be excellent tools for discovering the basics of home design — saving you time and money with your architect even if you don't design the entire home. If you find yourself with the time and skill to design the whole project, you can save significant money. Here are our favorite software choices available online or at any computer store:

- www.smartdraw.com
- Better Homes and Gardens Designer Suite
- 3D Home Architect

The same issues apply when designing your home using software as with store-bought plans — the need for foundational engineering and structural drawings. However, in this case you're now responsible for all the structural elements of the house construction. Make sure to find a structural engineer you can work with before heading down this path. You'll need to search the phone book or ask architects or local building departments for referrals to find a good structural engineer. Otherwise your new home may become your design nightmare.

Hiring a home designer

You can also choose from a growing number of talented home designers who aren't licensed architects. They offer you the possibility of significant cost savings in the design phase of building your new home. They don't have the architectural training or certification, so they bill at a lower rate than architects. These designers may draft the house design for you from scratch or help you determine materials. No standard for the services they provide exists, so you need to ask them what part of the process they will provide. Treat them as you would any architect. Investigate their credentials and experience. Discuss with them the differences in the services they will provide from architects and other designers and find out where the gaps exist. The best way to find these designers is in local newspapers and design magazines.

Many states limit designers from designing anything more than a bathroom, kitchen, or single-story remodel without the plans being approved by an engineer with an engineer's stamp before obtaining permits. In addition, a designer may not be aware of code complications in a more extensive project, creating more cost to fix the plans even with the engineer's assistance.

Placing the House on the Lot

Before you start designing your new home, you need to figure out how it will sit on the lot. Lot placement is important because it allows you to take advantage of views, topography, and amenities. Some lots may have special features or limitations that make this decision a simple one. Others that are large and flat may have limitless possibilities. We lay out some of the biggest considerations in this section.

Foundation issues

If your house is on a slope, then the engineers are going to give you limited choices in how to place the house on the lot. You'll have to follow very specific requirements for grading, piers, or other specialized foundations. If your topography is far from flat, you may want to consult an engineer early in the design process. You can do this through an architect or consult the phone book.

Constructing a foundation can be complex, and you need to discuss it with the architect or engineer during the house design process. To understand the specific process for hillside foundations, look in Chapter 12.

Which orientation is best? North, south, east, or west

There are no right or wrong answers for picking direction placement for your home — it's a matter of preference. Some people like the sun in the morning and some in the afternoon. If you're building in Seattle or Vancouver, the sun may not be a factor at all.

In North America, the sun shines from the south so a southern exposure means that the sun will shine on the front of your home for most of the day. A south-facing house also means your backyard may get little sun until midday, when the sun is high enough in the sky to shine over your house. Figure 5-1 shows the relative angles for exposure and house placement.

Too little sun can make the house seem dark and dank. Too much sun can be energy inefficient and weather the house prematurely. You need to consider other directional factors, such as wind, noise, and city lights, that can negatively impact your home. To decide on the right direction, ask yourself the following questions:

- What are sun patterns where I live?
- What is my preferred daily temperature?
- How bright do I want my home?
- How much sun do I need in each yard?

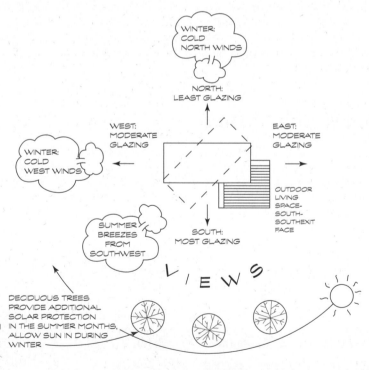

Figure 5-1: Exposure impacts placement.

Courtesy of Tecta Associates Architects, San Francisco

> ✔ Where does the wind normally come from?
>
> ✔ Where is the closest major city?
>
> ✔ Are there any noise issues in any of the directions?

After you have considered these questions, determining which direction will benefit your lifestyle and then designing doors, windows, and decks to face that optimum direction should be easier.

Taking advantage of natural elements

In some suburban neighborhoods, the most natural thing about your lot may be the hippie couple living next door. However many people building custom homes have some natural features that can add to their home's beauty. Following are some natural elements to consider in the design and placement of the home:

✔ **Foliage:** If your lot is in a rural area with plenty of natural landscaping, consider natural growth patterns for beauty and easy maintenance. High trees and bushes can afford you privacy; however, clearing tall brush and cutting back trees may give you unexpected views.

✔ **Mountains:** If you have a larger lot in a mountainous area, decide to be on top of the peak or shaded at the foothill. Or you may simply prefer to stare at the neighboring giant.

✔ **Rocks:** Small groups of stones or even large boulders can make for a dramatic effect depending on where you place your house. In some rural and mountainous areas, designers have built spectacular homes with boulders *in* the house making the home unique and saving the cost of demolition.

✔ **Trees:** Trees provide shade and beauty. They can also be a nuisance with dropping leaves and fruit. Like boulders, some old magnificent trees can be incorporated into the home's design.

✔ **View:** The right view can significantly increase your home's value and beauty. Try to optimize views for rooms where romance, relaxation, and entertainment occur.

✔ **Water:** People pay premiums to live by the ocean or near lakes and rivers. Take advantage of these aquascapes, but be wary of flooding issues by checking flood maps with your engineer.

Custom home projects tend to run smoother and cheaper when they're designed to take advantage of natural elements. Reconstructing landscape and waterways or removing huge trees and boulders can be costly and in some cases environmentally damaging. Look for ways to take advantage of what was naturally provided.

Nothing is as lovely as a tree

One very expensive cost for a new home can be trees. In new developments, the land is bare and the cost of transplanting a mature tree can be as much as $20,000. The expense of mature trees is the reason why it takes decades for neighborhoods to have tall trees. If you're lucky enough to have mature trees on your property, take advantage of them in your house placement. In some locations certain trees may be protected, such as live oaks in California (the state tree). You can get plenty of information about types of trees and how to protect them from the National Arbor Day Foundation (www. arborday.org).

When making significant changes to the elements, work with your engineer to create proper site drainage and insure the soil will remain stable with few erosion problems. Otherwise your house could slide down the hill or be buried by mudslides.

Planning the Size and Shape of Your Home

Even though building a custom home implies you do it your way, few people actually have enough money to put everything they want into their first project. Unless you recently won the lottery or invented the cure for the common cold (congratulations if you fall within either of those two categories!), you'll probably have some limitations on what you can build. Also, unless you're in a position where money has no bearing, you'll want the house to maintain its value and potentially appreciate.

So you now have three gods to appease:

- **Desire:** You want to build a house you want for your needs.
- **Taste:** You want the house to have aesthetic appeal, particularly to you.
- **Value:** You want to make sure the house is built in such a way to maintain its investment potential.

You must carefully consider all three of them in the design process.

Size matters — Figuring the right square footage

For reference, you'll often see square feet represented by a number with a symbol so that it looks like this: 2,000□. Three major factors dictate the appropriate square footage for your home:

- **You need to establish your family's needs.** For example, do you have elderly parents living with you who need a downstairs bedroom or does your wife want her own separate walk-in closet in the master bedroom? We address this topic more in-depth later in this chapter in "Ten general floor-plan considerations."

- **You need to adhere to zoning limitations or *covenants* (neighborhood guidelines).** Many design rules set limits on how big and how small of a house can be built on a particular lot. You may also encounter limits on the house's ground floor. These guidelines may impact other decisions such as the need for more than one story to meet your square footage needs. (See Chapter 3 for more on zoning regulations and covenants.)

- **Your need to keep in line with your budget.** In Chapter 2, we give you a method for determining a budget as well as a way to define *dollars per square foot.* You'll need to design a house that not only fits your family and the lot but also fits your budget as well. Many of these calculations go in circles, so start with the house you want and see if it fits based upon local estimates for building costs, which you can get by talking to a few contractors. If the going rate seems to be $100 per foot and your budget allows for $300,000, then a 3,500 square foot house won't work and you need to adjust your design.

The more square footage you build, the more the house will cost, so efficiency is important. At the same time, skimping on rooms can reduce utility and make for unpleasant living. Here are some minimum recommendations for typical room sizes to give you a general idea of what you need:

- Bedroom: 100 to 200 square feet

- Dining room: 100 to 300 square feet

- Family room: 300 to 800 square feet

- Full bathroom: 60 to 150 square feet

- Great room: 400 to 1,000 square feet

- Kitchen: 150 to 350 square feet

- Living room: 300 to 800 square feet

- Master bedroom: 200 to 600 square feet

- Staircases: 100 square feet per story

Calculate the total square feet of the rooms you have picked. You can figure on adding another 10 percent to account for hallways, cabinets, and closets. This total can give you a basis from which to start your estimate. Other factors to consider may include unfinished space like basements and garages. Garages can vary based upon size, but each car needs roughly 200 square feet. Generally, a basement matches the square footage of the first floor. If you're not going to finish the walls and flooring in your basement or garage, then you don't count it in your livable square footage, but you'll need to reconcile the cost when you get your estimates from contractors. You can estimate its cost now by multiplying the square footage by the dollars per square foot and dividing by 2.

Designing for resale — Create a house everyone wants to buy

When designing your home, you want to remember that a unique house can create difficulty even in a custom home — particularly thinking ahead to the future and your new home's resale value. For whatever reason, most people prefer houses that are familiar, functional, and comfortable, which means many people may find your home to be a nice place to visit but they wouldn't want to live there. If buyers aren't interested in your house, lenders will shy away as well making financing difficult. (We discuss the issues of marketability in greater detail in Chapter 7 and the lender's perspective in Chapter 9.)

Just because you want to design your house with resale in mind doesn't mean the house has to be generic. A number of proven theories in home design create functionality and appeal yet allow for uniqueness. Talk to local real estate agents and your architect about the expectations of most buyers in your neighborhood. You don't have to build your new home exactly for them, but at least you can consider them in your design decisions as you move through the process.

Exterior styles — Considering architecture examples

Some local guidelines require specific architecture styles for the neighborhood. Some design review committees may actually dictate the type of siding to be used and colors to be painted. We discuss these committees at length in Chapter 6. The key is to pick something that suits the neighborhood and your taste. You can choose from many examples of exteriors in plan books and on the Internet.

Knowing the size of the home isn't enough to get a handle on costs. Not all exterior designs cost the same. The more complex your exterior is, the more it will cost in framing (see Chapter 13). Architectural extras such as peaked roofs, dormers, and balconies can increase costs. Exterior materials have an impact on cost as well. If the neighborhood guidelines allow it, you'll have to choose between siding, stone, or stucco as well as roof material choices like slate, tile, or metal. We discuss these material choices extensively in Chapter 14. Do your research on these materials now by checking prices and discussing options with your architect.

Designing Your Home's Interior

Although design review committees may have a lot to say regarding your home's exterior, the interior choices totally belong to you. The floor plan determines where everything is located in your house. A well thought-out floor plan can make for a comfortable house whereas a bad floor plan can create problems and inconvenience.

Several components, such as doorframes and hallway passages, require you to make decisions about style, size, and location. For example, choosing an open feel in a house requires open passages and larger hallways whereas privacy needs may push you to opt for smaller cozier options. These decisions impact the feel of the house and, ultimately, your enjoyment of it.

Other interior choices on details such as corners and finish trim can add significant themes to the look and feel of your home's interior. You can find hundreds of interior ideas in the multitude of home magazines on the rack at the bookstore. *Interior Design* magazine is excellent for ideas and, of course, you can always find plenty of pictures in the classic *Architectural Digest* and on the Internet.

The best way we have found to search for ideas on the Web is to go to www.google.com, click on the "Images" button, and search terms such as *interior architecture* and *interior design*. You'll get hundreds of pictures to look at for ideas.

Ten general floor-plan considerations

Whether you're designing the house or using an architect, you need to be aware of elements of good home design. We lay out these elements in this section so you can incorporate them into your thought process while designing your home. Our experience shows that most design problems in the flow and functionality of the house result from not addressing one of the following areas:

- ✔ **Lifestyle:** To be comfortable and relaxed, your home has to fit your lifestyle. Not every home works for every family. Are you a family living in a formal style? If so, then maybe you need a formal living room and dining room. Do you gather around the kitchen? Do you entertain a lot? Then a home with a big family room open to the kitchen may work for you. Determine how you want to live and design a plan that fits the lifestyle you enjoy.

- ✔ **Foot traffic:** Try to project how people move through the house on a daily basis. Look for problems in the traffic patterns. Some problems may include issues like tight hallways and people crossing through work areas of the kitchen, formal areas, or TV-viewing areas. (You don't really want Little Johnny racing through your dinner party in his Spiderman tighty-whities to get to the bathroom.)

- ✔ **Noise:** This factor can be huge in multilevel houses. Remember that bedrooms are for sleeping, so any noise above, below, or next to a bedroom can make for a restless night. Consider carefully the placement of noise-generating rooms like the garage, home office, laundry room, and bathrooms. (For example, if you have a large family, you don't want to put the main bathroom by your bedroom. The last thing you want to hear is a flushing toilet all night.)

- ✔ **Storage:** You can never have enough storage space in any house. The longer you live somewhere, the more stuff you acquire. What you really need is plenty of useful places to put it. Create ample-size closets, pantries, and cabinet areas. And make sure your storage areas are convenient without being obtrusive.

- ✔ **Door placement:** Every room needs a door, but it needs to open in such a way that it doesn't bang against walls, obstruct other doorways, or block closets or windows. Consider the placement and opening space necessary for each door in the house.

- ✔ **Window placement:** What's the point of having a view if your windows don't take advantage of it? Other window considerations include privacy, not being blocked by doors, or looking out on the garbage area. Put plenty of thought into the size and placement of your windows to the world.

- ✔ **Accessibility:** Can you get to the outside from everywhere that makes sense? Are bedrooms and bathrooms easy to access from common areas? Make sure you can easily access important rooms without creating unnecessary obstacles.

- ✔ **Convenience:** We can think of nothing worse than having to traipse halfway across the house with food from the kitchen to serve in the formal dining room. Think about where you may be unloading your groceries or how to get the food from your barbecue. Bathroom placement is another major convenience consideration.

✔ **Utility:** Many homes have nooks and cubbies that serve no useful purpose. You'll pay to build any square footage whether or not it's useable space. Make sure all areas serve a purpose.

✔ **Future expansion:** Perhaps your new house is perfect for all your needs at this time, but someday your needs may change. Think about how you might expand the house should that occur.

Special considerations room by room

We realize many of you are hoping in this section for a detailed list of decisions on design in the various rooms of the house. But if we made all the design choices for you, then your new home wouldn't be custom.

In this process of custom building your home, all the decisions are yours – you're in charge. Our job is to share our insights on the questions. In this section we give you questions and suggestions to analyze when designing the rooms. You can use it like a checklist. Then you'll have a good basis for conversation with your architect or designer. For those of you designing your own house, use this section as a template for decision making.

A cook's tour — Kitchen elements

Isn't it funny how every party eventually ends up in the kitchen — often one of the smallest rooms in the house? Think about how much time you spend in the kitchen. Food is a critical part of family culture, and you want your kitchen to reflect it. Think about placement for breakfast eating areas. Where are people going to collect and connect? Although the kitchen is usually the most expensive room in the house, the kitchen also brings the best return on money spent in any home. For greater detail on kitchens, you can read *Kitchen Remodeling For Dummies* by Donald R. Prestly (Wiley). Here are our tips for assessing your basic kitchen needs.

Cabinets and counters

Your cabinetry sets your kitchen's tone as well as establishes its convenience. You have three major issues to consider with cabinetry and counters:

✔ **Layout:** You need to make kitchen layout choices in the floor plan including the specific layout of the counters and island. Put your time into the kitchen design early because making cabinetry and counter changes can be costly after materials have been ordered. Walk through as many kitchens as you can at open houses to see what works well for your lifestyle. The general rule is that you don't want to have to walk food over great distances during preparation. You also want to make sure you have adequate room for those appliances you've been lusting after as well as ventilation for your cooking needs.

✔ **Cosmetics:** The kitchen design is likely to stay the same for a long time. You can always change the look of other rooms by painting the walls, but you're less likely to change the kitchen cabinets in the near future for simple aesthetics. Cabinetry is expensive. Make sure you have chosen a style and color that will suit the house's style for decades to come. If you get bored easily, consider paint-grade cabinets so you can change the look just like walls.

When making a decision about countertops, you have many choices to consider, but it often comes down to a choice of beauty versus practicality. Tile can be cost effective and attractive, but grout can be difficult to clean. Granite or marble is gorgeous, but expensive and harder to maintain. You can choose from many suitable manmade alternatives such as Corian that will last almost as long as tile or stone. You can even choose from other surfaces, such as laminate, wood, zinc, copper, stainless steel, and even concrete, that have been used for utility and a unique look. Do your homework to determine which surface is best for you and your cooking style.

✔ **Size:** The size of your counters and cabinetry and how much storage space you'll need depends upon your cooking style and equipment. Many people like to display their pots in pot racks and others prefer a kitchen that hides everything cuisine related.

The choice of shelving and inserts requires much thought about your lifestyle and needs in the kitchen. Think about the way you like to cook in the kitchen. Make an inventory of all your cooking tools and machines. Then plan in advance and make a map of where they might live. If you're a kitchen gadget-hound that needs everything at your fingertips, then you'll want plenty of counter space and electrical sockets for your juicer, meat slicer, and George Foreman grill.

Feng shui — The art of balance

Feng shui has had a huge impact on design, especially when it comes to housing. Feng shui is an ancient Chinese system of philosophy, science, and art. Its purpose is to connect people with heaven and earth. It's based upon the interaction of the environment with energy and intention. The feng shui philosophy seeks to obtain a balance between opposites in the environment. So, for example, feng shui philosophies can determine room placement, window and door location, and so on. If you make a feng shui mistake, to maintain good feng shui, you may need to create fixes such as hanging coins or mirrors to remove imbalance.

Many people find feng shui suitable for creating their own interior design guidelines and even necessary for resale if building in a city with a large Asian American population. You can find out more about feng shui in Feng Shui For Dummies written by David Daniel Kennedy (Wiley).

Custom or prefab?

A big debate rages on about custom cabinetry versus units that are made in factories. The prefabricated companies claim quality control is far better and you can't beat the price. Today's technology allows for tremendous customization of prefab components suiting most situations adequately. However if you're looking to create something worthy of the museum of modern art, then you need a custom cabinet maker. Custom cabinetry can cost more than three times the amount of prefab, but they make better use of your kitchen space because they're designed to fit exactly. The good news? These artisans can create incredible pieces of curvy-grainy-spectacular-laminated art that will be the envy of kitchen guests for decades.

Appliances

You can probably expect your oven, stove, refrigerator, and dishwasher to last a minimum of ten years. More and more people are choosing cosmetically matching suites of appliances that are matched and installed. Home shows are the best place to find the appliances that will suit your budget and your cooking needs. Pick the largest and best you can afford. You can get great information and compare appliances at www.consumerreports.org.

Beware of deciding on industrial equipment. Some stoves designed for restaurant use have different power requirements and safety standards than consumer-designed equipment. Many companies, such as Viking, Wolf, and GE, make commercial-grade equipment designed for consumers.

Flooring

Aside from its appearance, the main consideration when selecting kitchen flooring is to remember that it gets the most traffic and requires the most cleaning. Carpet collects dirt and crumbs and so does tile grout. Smooth tile can be pretty, but every dish you ever drop will smash to smithereens. Gaining popularity are waterproof-composite floors such as Pergo or WilsonArt. These floors come in a variety of textures and styles impressively looking like slate or wood, making for a well-designed alternative to linoleum or vinyl. Many architects believe that the best surfaces for kitchens are wood or stone.

Bathroom considerations

After the kitchen, bathrooms are the next most expensive rooms in the house. The labor necessary for all that electrical, plumbing, and tile work adds up quickly. Plus you only have to create one kitchen, whereas you may have multiple bathrooms. Each one can rival the Taj Mahal if you want. You can find specific details on designing good bathrooms in *Bathroom Remodeling For*

Dummies by Gene Hamilton and Katie Hamilton (Wiley). Here are the major considerations for the bathroom:

- ✔ **Size:** A full bath has a sink, a toilet, and a bathtub/shower. For a three-quarters bath, take out the tub; a room with only a sink and toilet is a half bath or powder room. Decide what is necessary for each designated area. Master suites and guest live-in areas generally require more space with full amenities.

- ✔ **Surface:** Tile has the cost attractiveness as well as decorative versatility, but grout can be difficult to keep clean. If using pedestal sinks, you can save on counter work. Complete prefabricated cabinet and counter units are also available for less formal cost-effective bathrooms.

- ✔ **Ventilation:** Most exhaust fans are installed for code but serve little purpose. Decide on your most desired form of ventilation. An open window still serves as the most popular and efficient. Be alert to sightlines for privacy from neighbors.

- ✔ **Luxury:** Big sweeping claw-foot tubs, Jacuzzi tubs, built-in saunas, steam showers, and the like are available to make your master bathroom your slice of heaven. Many of these amenities need to be installed early in the process and require high maintenance, which can be annoying and costly. Make these choices early and research to see which pieces appear relaxing but are really more trouble than they're worth.

Bedrooms and home offices

Most people don't designate between guestrooms, den, office, or workout rooms because these rooms change based upon the usage of the family living there. The most versatile designs give these rooms easy access to bathrooms and equal appeal as the design allows.

The master bedroom is your reward for paying for this project. You want it to have plenty of room for relaxation with a great view for those romantic moments. A touch of privacy is desirable, so placement of the master bedroom away from other bedrooms and heavy traffic areas helps provide seclusion. Also, plenty of room for closet space is a required necessity for shopaholics.

Aside from the master bedroom, the other bedrooms need to be designed for optimum utility. Take advantage of light and views where you can and make sure each room has adequate storage space. Each wall needs to have at least two electrical sockets to accommodate technology.

Closets can gain greater clothes capacity through closet organizers. A wide variety of companies today manufacture design systems and materials for creating efficient closets that hold significant amounts of clothes, shoes, and stuff. To get an idea of available options, check out www.california closets.com or for you do-it-yourselfers, try www.closetmaid.com.

Dining and entertaining

Dining rooms and family rooms often center around food and need to have reasonable access to the kitchen. Decide whether a television is a critical part of your food time; otherwise you may find yourself eating on TV trays in your family room. If you want a home theater, understand that it'll have its own special needs in terms of acoustics and technology. You can start that research with *Home Theater For Dummies* by Danny Briere and Pat Hurley (Wiley).

The biggest mistake we have seen in entertaining rooms is people wasting money on built-in furniture. For storage, your needs will change — and so will the furniture. Built-in cabinets in dining rooms can go out of style or create furniture placement problems. Building furniture around TVs and stereos has proven to be a bigger waste of money as the technology changes make TV styles obsolete every seven to ten years. For example, yesterday's bulky projection TV has given way to the compact, wall-mounted flat screen, making large cabinets useless and cumbersome.

Running water will be an important consideration for entertaining. Any rooms with a wet bar need running water, drainage, and power for a dishwasher and the all-important blender for daiquiris and margaritas. Additional cabinetry may be necessary as well.

What's in a garage

Some people may consider the garage as only a home for their vehicles. However, other people view their garage as a workshop and storage unit. Your garage can serve all these purposes with a little planning.

When designing your garage, think about storage access as well as the space. You want to get to everything while avoiding obstacle courses or throwing out your back. Consider dumbwaiters for storage above. If your space includes a workshop, take into account ventilation and noise. You may add extra fans and insulate with soundproofing, which adds to your comfort in extreme weather. Make these choices early on so you can plan for power and water needs accordingly.

The Devil Is in the Details

Earlier in this chapter we discuss basics for the house, but really the little details will make this house something to cherish. For those of you who love detail work, these projects are just what you're looking for to personalize your new home. If you don't love detail work, take some time to get organized because you can't leave out anything. If you're completely detail-challenged, you can pay an architect or a designer to help you through this process.

Many people choose to leave the details to the contractor, but doing so often leads to misunderstandings on types of materials and costs late in the process. Avoid the headache. Make the decisions in advance and communicate with your contractor what you want.

Materials, hardware, fixtures, and finishes

Check out Table 5-1, which contains a comprehensive list of items that you need to consider for your new home and different questions to ask yourself. Chapters 12 and 13 outline the installation for most of these materials as well as certain advantages and disadvantages. Use this table as a shopping list for when you're estimating your costs.

Table 5-1	Important Details to Consider
Specific Items	*Questions to Answer*
Appliances	What brands? Do you want free-standing or built-in?
Baseboard	What type of wood? What kind of finish?
Carpet	What type do you want? Wool or synthetic? What color? What kind of pad? How thick do you want the pad?
Crown molding	Do you want it? What type? What kind of finish?
Doors	What style do you want? What type? What finish? How many?
Door handles	What style and color do you want? How many?
Door hinges	What type and style do you like? What color? How many do you need?
Eaves	What type and finish do you want?
Exterior facade	What color and style do you want? What type of material do you like?
Exterior trims	What type and finish do you want to match your exterior facade?
Faucets	How many do you want? What styles? Do you want any outdoors?
Fireplace	What type of face and mantel do you want? Do you want a hearth? If so, what type? Will it be gas, electric, or wood burning?
Floor tile	What style and color do you prefer? What color do you want for the grout?
Front door	Do you want glass or solid? What type of material, fiberglass, or wood? What color? Do you want a screen door?

Specific Items	Questions to Answer
Handrails	What type of wood do you want? What type of finish?
Hardwood floors	What style do you like? What kind of thickness and width? What color stain do you prefer?
Heating and air conditioning	Will it be a forced air system? How many pump units? How many tons of cooling? Will you have any radiant floor heating? Will you use Zone heating and cooling for efficiency?
Insulation	Will it be rolled insulation or blown? What rating will it be?
Interior walls	What type of materials? What kind of finish?
Lighting fixtures	How many do you want? What types?
Roof	What type of shingles do you want? Do you want a flat or pitched roof?
Wainscot	Do you want it? What types of finish and wood do you prefer?
Wall tiles	Do you want them? What type (decorative, monochromatic, accents)?
Windows	What thickness do you want? What type (metal, wood, or vinyl)?

Make all your decisions now — Allowance is a dirty word

Be prepared to put a lot of time into the material selection process. You'll have to make all those decisions at some point. It's never a question of how much time you'll have to spend on this shopping process; it's merely a matter of when you'll do it. Our recommendation is to select materials as early in the process as possible.

Many people spend less time on the small details during the design process and defer them until their home's basics are erected. They take their plans to the contractors who bid based upon estimates for the finish materials they call *allowances.* Doing so is a recipe for disaster. Contractors make their own decisions about the quality of materials you want and their assumptions may not be accurate. Also, you may not like what they have chosen and it may be too late to get what you really want.

In our experience, putting off the shopping for materials until the end of the project is the No. 1 reason for projects going over budget! Spend time at the beginning to make the decisions or pay your architect extra money if you're using one. You need to pick out every hinge, fixture, and appliance before you get price estimates for your project. This way you insure that all bids from subcontractors are equal, and you can be sure of the availability of materials. Doing so also removes surprises and gives you the most accurate financial picture.

The Internet is an awesome place to find materials prices and even unique hardware. Just go to www.google.com and type in "hinges" or "doorknobs" (or whatever you're looking for), and you'll be delighted with the many choices available. You can even buy antique lamps and hardware in a cost-effective manner at auction sites such as www.ebay.com! For a more hands-on experience, you can go to showrooms, such as Ikea (www.ikea.com) or Home Depot Expo (www.expo.com), to see kitchens and bathrooms and get a feel for functionality.

Energy efficiency — Saving the earth (and your money!)

We can think of several areas in your home that you can enhance to conserve energy and be more environmentally friendly. We list a few here and provide additional resources in Chapter 22.

- **Doors and windows:** Today, advancements with double-pane windows and gas-filled panels reduce the ability of glass to transfer energy. Check out companies like Marvin Windows at www.marvin.com, Pella at www.pella.com, and Anderson at www.andersonwindows.com for the latest achievements.

- **Insulation:** A tight house is an efficient house (but keep in mind that a house that is too airtight can be an unhealthy house). Houses need proper ventilation so the air stays fresh and you don't breathe your own carbon dioxide. You can research or discuss with your architect methods and materials that provide maximum insulation at a reasonable cost. In extreme environments, insulation will be the No. 1 factor for energy savings.

- **Energy-efficient appliances:** Many manufacturers make lines that focus specifically on energy savings. Many local utility companies offer rebates for choosing appliances with lower energy ratings.

- **Solar power:** Many people have saved money by supplementing their energy with solar panels. The technology has improved since people started using solar power in the 1970s. Panels have become smaller and lower profile. Many options are available through sources on the Internet.

> ✔ **Heating, ventilating, and air-conditioning (HVAC) system:** Over-estimating the cost of HVAC happens often because everyone wants a more efficient air system, a reduction in noise, and comfort, especially when the weather is extreme. Larger systems may cost more, but you typically can make up the extra cost in savings when the energy bill comes around.

Considering technology options

Technology has never been as much a part of individuals' every day lives as it is today. The Internet is a regular part of home life, and more and more people can work at home because of it. Custom homebuilders often tend to want to add every new piece of technology offered. Check out *Smart Homes For Dummies* by Danny Briere and Pat Hurley (Wiley) for more details about all the different choices.

Kevin has seen the pros and cons of working with house technology over the last several years, and he has a few pointers.

The less technology-specific the home the better

Technology changes faster and faster today. Kevin has clients who only five years ago spent a fortune to run state-of-the-art Cat-5 computer wire throughout their homes. Today, wireless routers render the wire obsolete.

Design your house to accommodate any new technology by providing tubing (conduit) and cubbies that give you general access points to rooms in the house. Mark the access points clearly so you can always add things where you want them.

Make sure you have ongoing support

Technology companies come and go. Sometimes the most innovative go up in a ball of fire the fastest. Whatever technology you install in the house needs to be maintained and serviced. If the manufacturer goes out of business, your technology may be as useful as that 8-track tape player in your attic.

Watch the budget

Your house project may take years, and new features that are better than what you install will be available. Buy only what you're truly likely to use. Kevin has one client that spent more than $350,000 making his house a smart house. By the time the three-year project was finished, most of the technology could have been installed for a mere $35,000. Talk about one unhappy client!

Chapter 6

Engineering and Plan Approval: Bureaucracy Made Somewhat Easy

Wouldn't life be great if you could simply sketch your house on the back of a napkin and the contractor would magically build exactly what you had in mind? Whether designing a log, stick built, or timber frame home, sadly the process is a bit more complicated; your project requires outside expertise and approvals. In this chapter we take the mystery out of looking at blueprints and plans. We talk about engineering the working drawings. We also explain the design review process and discuss acquiring and paying for permits.

Understanding Plans and Blueprints

If you have ever built model cars or airplanes, you know how important the instructions are. Nothing gets you into trouble faster than trying to assemble that model by looking only at the picture on the box.

Obviously, having instructions when building a new home is essential. However, because a house is a complex structure made of many different systems, your instructions (or plans) need to include many different drawings. A typical set of plans will include 30 to 50 pages of specific instructions on how to build your house. The plans first include a set of preliminary designs or *prelims*. After these prelims are approved, the engineer prepares the working drawings for constructing the house. (See "Working drawings: The how-to-build-it papers" later in this chapter for more on working drawings.)

Why are blueprints blue?

Blueprint is a long-surviving term, more than 150 years old, that comes from the fact that reproductions of plans for construction were always blue with white lines representing the drawings and words on the page. The construction industry needed a way to make exact replicas of large drawings with exact measurements because the corner Kinko's was unavailable. Architects and draftsmen first drew their plans on tracing paper. The translucent paper was placed on light-sensitive chemically reactive paper and soaked in chemicals that turned all the light exposed areas blue and left the lines white. This system was a cost-effective method for creating multiple sets of perfect duplicates suitable for construction. Later, the process was reversed to create blue-line prints where the lines are blue and the paper is white. Although blueprints and blue-line prints are still used today, most architects now use Computer Aided Design (CAD) systems and simply print plans on large printers. The term blueprint has stuck and now means any sort of master plan.

Architects and engineers often draw plans in quarter-inch scale, meaning that each ¼ inch on paper represents 1 foot in real life. You can read the drawing measurements easily with a ruler by measuring any line and dividing the number of inches by 4 to understand how many feet long any straight line will be in real life.

Prelims — Floor plans, site plans, and elevations

The first designs will be rough sketches and drafts drawn by an architect or designer. They may include scratch drawings and renderings, which are an artist's version of what the house may look like. If you buy plans from a book or online as discussed in Chapter 5, you can skip the rough sketch phase of the process. Plan software also discussed in Chapter 5 provides a neat way to try different floor plans with ease.

As soon as you and your architect, if you're using one, have made some basic decisions on style and size, the architect will draft preliminary drawings. These drawings are necessary to show the house in three dimensions. The prelims will be used primarily for making initial decisions, such as room placement and size, with your architect, as well as preparing for the initial design approval process. (For the nuts and bolts on the design approval process, check out "Submitting Your Prelims for Approval" later in this chapter.) Creating these prelims is an ongoing process of reviewing drawings and making changes. If you're buying plans, the plan company provides you with the prelims. If you're using a software program, the prelim creation is your responsibility.

These *prelims* consist primarily of three basic elements:

- ✔ **Floor plans (see Figure 6-1):** Each floor of the house has a layout that shows the following:

  - The location of each room

  - The placement of each door and window

  - The location of other amenities, such as stairs, fireplaces, closets, and major fixtures such as kitchen cabinets and showers

- ✔ **Site plan:** The site plan shows how the house and other buildings such as the garage will sit on your lot (see Figure 6-2). The site plan explains the position of the house and the direction it faces. It also specifies how far it sits from the street and neighbors' houses based upon the required setbacks discussed in Chapter 3.

- ✔ **Elevations:** The elevation drawings illustrate what your house's exterior will look like from the ground up on each side (see Figure 6-3). Most elevation sets show the house from all four directions. The plans illustrate exterior windows and doors as well as any ornamentation in the design. From these plans, you can measure the height of the structure and its elements.

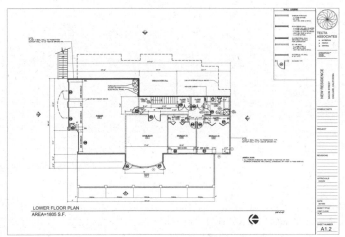

Figure 6-1:
Example of
a floor plan.

Courtesy of Tecta Associates Architects, San Francisco

Working drawings: The how-to-build-it papers

After the prelims have been finalized and approved, the architect and engineer create working drawings. *Working drawings* are a series of individual papers giving explicit instructions on how to build the house. They give you every detail for construction including where to put the plugs and switches as well as the number of rafters in your roof. Furthermore, the working drawings include all the technical specifications and requirements for engineering and compliance with building codes. Each of the systems in the house is specified in the working drawings.

Figure 6-2:
Example of
a site plan.

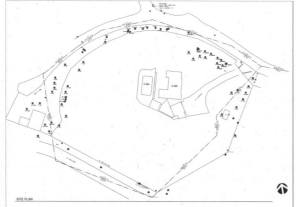

Courtesy of Tecta Associates Architects, San Francisco

Figure 6-3:
Example of
an elevation
drawing.

ENTRY ELEVATION/ EAST

Courtesy of Tecta Associates Architects, San Francisco

In addition to the floor plans, site plan, and elevations, a typical set of working drawings has individual drawings for each structural system of the house. These individual working drawings

✔ Provide all the technical specifications necessary for contractors and subs to bid on your project. Each different section goes out to a different craftsman so they can determine the time and materials necessary to complete the project.

✔ May include some variations if you're working with a design-build firm like we discuss in Chapter 5. Because the same firm will be handling both the design and building process, it may combine or reorder some of the technical pages to fit its process.

✔ May contain other pages that detail specific construction of parts that require extra detail such as unique staircases or particular architectural elements.

✔ May also include pages specifying energy calculations where required by the building department.

Working drawings generally include the following documents drawn in equal scale:

✔ **Architectural plans:** Site plan, floor plan, elevations, cross sections, wall sections, schedules of materials, and details

✔ **Civil plans:** Site plan, grading plan, and details

✔ **Electrical plan:** Outlets, switches, and lighting plans (see an example in Figure 6-4)

✔ **Landscape plan:** Landscaping layout, irrigation plans, schedules, and details

✔ **Mechanical plans:** Furnace and ducting plans and details

✔ **Plumbing plan:** Plumbing riser plans and isometrics, and details

✔ **Structural plans:** Foundation plan, framing plan, cross sections, and details

Building codes and ABCs

Building codes are rules and guidelines that specify how construction should be completed so that buildings will be safe. They address issues such as the requirements for electrical wiring, the size of pipes, how far to place framing posts, and so on. Unfortunately, no one standard building code exists for the entire country (or for the entire world), although it would certainly be easier if there were. Most municipalities have adopted a regional code and then created variations as required for their area.

In the United States, the basic codes come from the following published codes:

✔ **The Uniform Building Code (UBC):** Widely used in the West.

✔ **The Standard Building Code (SBC):** Widely used in the South.

✔ **The National Building Code (NBC):** Until recently, widely used by everyone not using the UBC or SBC.

✔ **The International Building Code (IBC):** The latest code, being adopted by everyone in all regions.

Your contractor, subs, architect, and engineer need to be knowledgeable with the codes for your area. Some regions, such as California, have more stringent code requirements than others due to seismic or other environmental issues. If you want to find out more about building codes or check a specific code, try www.codecheck.com.

You can see an example of how these system drawings are isolated in Figure 6-5 so each sub can focus on his or her work.

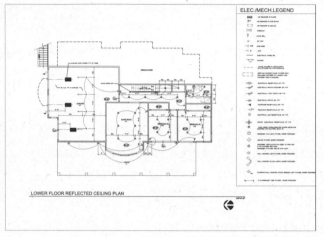

Figure 6-4:
This electrical plan shows the electrician where to wire the house for sockets, switches, and junction boxes.

Courtesy of Tecta Associates Architects, San Francisco

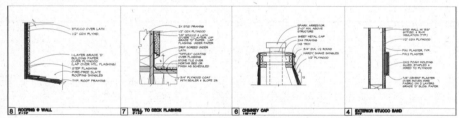

Figure 6-5:
Each box explains to the sub specifically how to construct an individual section of the house.

Courtesy of Tecta Associates Architects, San Francisco

Working with the Building and Planning Departments

Many of the nightmare stories you may hear about custom homes stem from dealing with local government during the plan approval process. The local county or city has to approve your designs and make sure the plans fit with their rules and regulations in a process called *plan check*. You have to provide to the local government offices all your plans and anything else they may ask for, which varies with every department in every municipality.

If everything goes smoothly, then you can be ready to break ground in three to six months from the time you first call your local planning office. (Of course, this time frame varies widely depending upon where you live.) If you hit a snag, then all bets are off. Kevin has clients who have battled building and planning departments for more than a year. The best way to prevent problems is to work with experienced architects, engineers, and contractors (if involved at this time) that know the department you're submitting to. These professionals can use their relationships and knowledge of the local government inner workings to chart the fastest, smoothest course for approvals and save you from making costly mistakes.

Here are additional pointers to remember to make the process of working with the building department go smoothly, especially if you have a problem:

- **Keep communicating.** First and foremost, keep the lines of communication open. The permitting process is all about passing information back and forth. Ask a lot of questions so you're sure you understand how everything works.

- **Be complete.** Most building departments hand out or post online the information required to apply for a permit. Have all your information together in a nice neat package. Make sure it's complete when you turn it in. If you piecemeal the process, you'll frustrate the clerks, inspectors, or plan checkers and they won't be able to make informed decisions — possibly leading to costly delays.

- **Deal with one person.** One helpful person can make all the difference in a building or planning department. Dealing with the same person can keep you from having to explain your situation again and again. Find a person that you can work with. If you respect this person and give him or her a pleasant experience, then he or she is more likely to give you one.

- **Have a single point of contact on your end.** Plan-checkers, clerks, and inspectors get frustrated and confused by getting what can be conflicting information from the architect, the contractor, and the client. Pick a contact on your side, keep in touch, and trust in your contact. Most people choose to make the architect or contractor the contact because they usually have longstanding relationships with the local employees.

- **Be persistent.** Most building and planning departments are underfunded and understaffed. They're busy, and there is always a bigger problem to take your place. Don't be afraid to call regularly to get the response you want. Be careful not to pester needlessly however. Pick your battles, but when there is something you really need, press the issue. Letting them know the realistic timelines at stake helps so they can set priorities in their workload.

✔ **Don't be intimidated or intimidating.** These people are civil servants; they're there to help you and want to do so. Don't be afraid to be ignorant of the building process, and don't be afraid to ask questions. At the same time, treat everyone in your building department with courtesy. Even if you hit a roadblock, you'll get further in the process with calm discussion than with angry theatrics. Smiling and saying thank you goes a long way in reminding people that you're human. A friendly tasteful joke once in a while may make that civil servant the inside friend you need.

Submitting Your Prelims for Approval

Before you spend thousands of dollars engineering plans and creating working drawings, you'll want to get approvals for the preliminary design of your house.

Generally, in rural areas, the design approval process is simple and has minimal limitations. In rural areas, houses are typically separated by vast acreage and the county is seldom concerned about what your house looks like as long as it meets the safety and building code requirements.

But if you plan to build in a higher density neighborhood or planned community, the guidelines can be strict and the design-approval process exhausting, especially if you're looking to bend the rules. Keep reading for info about the design review process.

Addressing grading, septic, and well issues

Before you submit your prelims for approval, the county may require special separate permits for specific items such as grading, well, and septic systems. These approvals and permits may need to be handled by you, your architect, or your contractor before the house plans are completed based upon the needs of the lot. You may encounter some restrictions with grading, well, and septic systems. For example, the building department may only issue grading permits at certain times of the year to prevent erosion when grading a hillside prone to geological issues.

Find out about any grading restrictions in your area before starting the permitting process. By taking time-sensitive issues into account in the beginning, you can avoid stopping and starting your project, which can create financing and labor problems down the line.

Wells and septic systems need to meet county standards for habitability as well as environmental concerns. The county may have minimums for water pressure allowable for the size of the house. A similar issue can exist for septic systems. If the soil doesn't support a standard system, the county may require a more expensive engineered system or restrict the size of the house.

Exploring these issues first with your architect and other professionals who can certify wells and septic tanks can save you time and money. Otherwise you may run into a brick wall on your approval and have to redesign the house from scratch or — worse — find out you can't build at all.

Understanding design guidelines

Have you ever wondered why all the houses in a neighborhood seem to have a similar look and feel? The similar look and feel isn't by chance. Neighborhoods, cities, and counties often develop specific guidelines that govern what you can build and how you can build it. Furthermore, other agencies such as the Coastal Commission may also have to approve your design if you're building near the coast. You can obtain these published planning and building department regulations from your local planning and building departments.

You also need to be aware of other guidelines that were put in place when your lot or neighborhood was created. You can find some of these rules in the covenants, conditions, and restrictions (CC&Rs). You should have received a set of the CC&Rs when you bought the lot. If you didn't, you can ask the title company from your purchase to get you a copy for free. Subdivisions older than 50 years likely have a minimal set of rules. They may not have CC&Rs that are pertinent.

Newer subdivisions are usually developed with specific themes. Seaside areas or golf course developments may have very restrictive guidelines to make sure you're staying within the theme of the development. No one in a planned Spanish-style development wants to drive by a giant Cape Cod house everyday. You can obtain these rules from the homeowner's association (HOA). Depending upon your neighborhood, the guidelines may include the following:

- Architecture style
- Drainage
- Environmental issues
- Exterior finish materials
- Height of the house
- Landscaping restrictions

- Minimum and maximum size of the house
- Paint color
- Types of the following building materials:
  - Doors
  - Roof
  - Windows

Developers and HOAs establish guidelines to ensure that no one house is imposing upon the pleasure of any other homeowner in the neighborhood. Most likely, after you have successfully built your house, you too will adopt a NIMBY attitude (which stands for Not In My Back Yard) and will become a vigorous defender of your neighborhood's guidelines.

Requesting variances and exceptions — Don't be Don Quixote

Some smaller cities and neighborhoods have volunteer design review committees, whose members may be appointed or elected by the community. Although such committees may have a government employee involved or in attendance, most are dominated by residents concerned with the preservation of their neighborhood's particular aesthetic feel. Design review committee members are passionate about determining what other people's houses must look like, so you'll need approval from this committee to build your new home. If you're lucky and careful, the process can be short and sweet; if you're not lucky — or worse, careless — the process can take far longer and be more harrowing than the building of the entire home itself.

Because you're building a custom home, you have already made a statement that you're dissatisfied with the homes already available in your neighborhood or community and want something more unique to suit your taste. As a result, design review isn't likely to be a picnic. For design review, you'll submit your preliminary plans to the design review board, neighborhood association, and/or planning department to get approval for the basic design of the house. They will review the plans and return them to you with a list of everything they don't like. *Subjective* guidelines such as style and colors may provide for negotiation for an *exception*. Objective rules such as exceeding height limits likely require a *variance* to the guidelines.

The best way to deal with variances is to design a house that doesn't require them. Working with an architect familiar with the community's regulations may help. If you're looking at too many restrictions for what you want to build, you may want to consider a different piece of land entirely.

No one has the right to a variance or exception. You can request it but it may be turned down. Some people choose to fight restrictions legally by showing evidence that similar variances have been granted in the past. If you undertake such a battle, you may win but at a great expense of time and money. Remember, even if you take your case to court, there's no guarantee you'll win. Furthermore, you may ultimately win a long drawn-out design review battle and end up losing the war. Don't forget that the design review board is comprised of your neighbors. A protracted battle filled with hate and lawsuits can make for a rather hostile housewarming when it's time to move in.

Not so fast — Acquiring neighbor approval

Just because the local design review association or planning department says they like the house doesn't necessarily mean you're home free. In many cities you still have to get neighbor approval before presenting your project to the design review board or planning department. Getting neighbor approval most often occurs when you're building in a long-established neighborhood, but not necessarily. This neighbor approval process gives you the opportunity to inform your neighbors and prepare them for the new addition to the neighborhood before you submit your plans. (The city will tell you if you need neighborhood approval.)

As a part of seeking this approval, you may be required to erect *story poles*. Story poles are wooden boards that outline the perimeter and height of the house as shown in Figure 6-6. These poles are required to stay in place for a designated period of time while your soon-to-be neighbors assess if your house will block their sun exposure, lake views, and so forth. Your neighbors will then be given the opportunity to challenge the approval of your building plans, which can turn into an unpleasant experience.

The earlier you establish a friendly, working relationship with your neighbors, the better your overall custom home experience will be. Just by planning ahead, you can reduce the stress level when submitting your plans and possibly start to develop good relationships with your new neighbors. Consider the following:

- ✔ **Take into account your neighbors' sight lines and exposure before you commit to the design.** Ask if you can look through their windows to see the impact on their property.

- ✔ **Put yourself in their position.** What would you think of living next door to your planned home?

✔ **Get their feedback and support for your project.** Kevin has one client who threw a wine and cheese party for the neighbors so they could come and see the plans before submitting. He had rousing support at design review. This process has been a favorite of architects he has worked with as well.

Figure 6-6:
Story poles are required in some cases to let neighbors know how your house will affect sight lines and aesthetics.

Courtesy of John Stetson

Gathering the Permits You Need

After the prelims have been approved, you can start moving down the permitting path. Don't worry; this process is a lot more cut-and-dry than the design review. The building permits are there to make sure that your final plan meets with the minimum standards required by your local city or county. The rules may be set for reasons of safety, logistics, or for environmental or other reasons. (No one has come up with a code to restrict teenagers to their rooms yet, but who knows, maybe soon.)

You or your architect will hire engineers to make sure you're meeting the needs of your local government and help create the working drawings. The permitting process isn't always a short one. If everything goes smoothly, you may get through it in a few months. Some processes can take more than a year with complications or bureaucratic difficulties.

Many building departments have a preliminary review process. For a fee, they look at all your working drawings after design review or before you formally submit the information to the planning department. They'll let you know if they notice anything questionable or potentially against code so you can fix it first. Utilizing the preliminary review process can get you through the formal process fast and easy.

Submitting and revising the working drawings

After the designs are approved and the working drawings finished, you, your architect, or your contractor (if involved at this point) submit the working drawings to the building department with appropriate fees required for application. These fees get the ball rolling.

The plan checkers scrutinize the working drawings to make sure they meet all the local codes and requirements. The plan checker marks the working drawings with red pencil for correction or in rare cases turns the plans down for permit. (If turned down flat, you'll need to have the plans re-engineered.)

You, your architect, or your contractor, if you've hired one yet, can pick up the marked plans at the building department. The architect and engineer make the requested changes to the working drawings and resubmit the plans to the building department.

If the plan checker still isn't satisfied, he'll red pencil the plans again and request they be fixed again. This process continues until the plan checker is fully satisfied, at which point the plans are officially approved.

In some areas, you may need to go through this process more than once. The plans may require review by a local city department and a county department. Different local governments have different jurisdictions for code, so they'll dictate their own approval process. Discuss the plan check process with your architect and engineer so you're clear on the steps and timing.

Picking up permits and paying the fees

After the building department has fully approved your plans, you or your contractor, if he's involved at this point, can pick up the permits.

You'll write a large check for all your permits and the remaining fees and pick up the permits so you can break ground. The costs vary widely from area to area, as do the names for the fees, but you can expect your permits and fees to cost between $3 and $10 per square foot of building area. You'll have to pay these fees in full before you're allowed to begin building on the property. Often these fees cover general expenses for the neighborhood's local infrastructure as well as overhead for the local government. (Remember, the city or county has to cover every cost and support the numerous staff members necessary to run their departments.)

The following list contains many of the fees you can expect to pay:

- Appeal fee
- Building permit
- Design review fee
- Drainage study fee
- Grading permit
- Land use permit
- Parks and recreation fee
- School fee
- Tree permit
- Variance fee
- Walkway fee

Although the cost of these various fees shouldn't be deal breakers for you, be sure you have sufficient funds budgeted for them and that you're prepared to pay when the time comes. Even though you may have to pay these fees out of pocket, you may be able to reimburse yourself through your construction loan if you included them in your budget. See Chapter 9 for more details on how these fees fit into the construction loan budget. These fees can often be reimbursed immediately by showing the bank the receipts. We talk about reimbursing these fees in Chapter 10.

Part II

All You Need Is Dough: Financing Your Custom Home

The 5th Wave   By Rich Tennant

"I'm well aware that we ask for a lot from our construction loan applicants, Mr. Harvey. However, sarcasm is rarely required."

In this part . . .

Your custom home project needs plenty of money. Either you have the cash or you don't! In this part, we explain the need for cash and how to determine if a lender is necessary. We walk you through the entire construction lending process including getting approved and picking the right loan. We also show how you get your money from the lender during the building process.

Chapter 7

Cash Is King: Using Debt to Your Advantage

*T*his chapter may be one of the most challenging to accept, but within its pages, you can find some of the most important concepts in this book. A construction project lives and dies on the availability of funds. Without access to cash when it's needed, a construction project will quickly come to a grinding halt. Most people want to be fiscally conservative when dealing with a large financial project like building a home. The challenge, however, is in understanding exactly what fiscal conservatism is really all about.

To make good decisions with your finances, spend some serious time educating yourself. No need to bury your head in the sand from the fear of numbers; you can access all kinds of information on the Internet and through the help of financial professionals. The more you find out about financial management, the better equipped you'll be to make the right decision for your situation.

In this chapter, we explain why having adequate cash on hand is so important. We also help you put borrowing in the proper perspective, and explain concepts that can help you protect your investment and manage your cash. Finally, we give some tips for deferring major financial decisions until the end of the project, when you have the information you'll need to make a good decision.

Accepting the Need for Cash, Cash, and More Cash

You may be wondering why cash is so important if you're already planning to get a construction loan and not relying on savings to fund building your new home. The answer is simple: Whether or not you borrow money, your custom home project will suck up cash like a vacuum cleaner. If you have enough money, you'll be fine, but run out of cash and you'll be in a big time world of hurt.

A *construction loan* does cover a significant portion of the funds to build your home, but you'll also face some significant restrictions and procedures for getting that money, so having extra cash on hand will keep the project moving. Not only that, but you need some cash before your loan is in place to cover expenses like permits, and after it's complete in order to pay for moving in and landscaping. In addition to your construction loan, you may end up needing as much as 40 percent of your total budget in cash to make the project work. We explain how construction loans work in Chapter 8 and discuss how you get your money from construction loans in detail in Chapter 10.

If you're still not convinced, consider the following reasons why extra cash is necessary even if you're financing your project through a bank:

- **You need a down payment for your land.** Chances are, when you purchase your land, you'll need to make a down payment and pay for closing costs. See Chapter 3 for details.

- **You need to make the loan payments and pay taxes on your land until the construction loan is in place.** It may take several months from the time you buy your land before you're ready to build and get a construction loan. You'll have to make payments on your land loan and pay property taxes while carrying the payment on your current home.

- **You need to pay the soft costs.** You have to pay for *soft costs,* such as the plans, engineering, and many fees for permits before you can fund your construction loan. See Chapter 6 for details about soft costs.

- **You need cash to close the construction loan.** Your construction loan may not be big enough to cover all your costs for the project. Also, the bank will want to see money in the bank beyond what is necessary for the project for qualifying. See Chapters 8 and 9 for details.

- **You have to pay the monthly costs during construction.** Your loan may have an *interest reserve* (or a pool of money set aside) to cover the construction loan payments as described in Chapter 8, but you still have to pay for property taxes and homeowners' fees as well as the house payments where you're currently living.

✔ **You need to fund the work between the construction *draws* (fundings from the bank).** The construction lender won't give you loan proceeds for a particular part of your project until the work is done. You'll have to put the money upfront, and be reimbursed by your construction lender with the loan proceeds later. See Chapter 10 for details.

✔ **You want to upgrade as you go.** You may want to improve certain items as you see the house come together, for example, substituting granite kitchen countertops in place of the Formica ones you originally specified. Because you can't increase the construction loan after you fund it, you have to pay for the upgrades out of pocket.

✔ **The project costs more than you budgeted.** Do you want some sobering news? Approximately 95 percent of custom home projects go over budget. (Can you say ouch?) And when a custom home project goes over budget, guess who pays? You. Although you may be the lucky 1 in 20 who doesn't go over budget, do you really want to take the chance with the largest investment you'll probably ever make in your entire life? We didn't think so!

✔ **Landscaping, decorating, furnishing, and moving expenses.** These items probably aren't included in your construction loan budget. Even if landscaping and hardscaping *are* included, the decorating of a new home can cost a pretty penny.

✔ **Something goes wrong.** Always remember Murphy's Law: If something can go wrong, it will go wrong. Anything from your contractor winning the lottery and abandoning your project to August snowstorms in Arizona can unexpectedly create the need for more extra cash. Just remember, as Yogi Berra said, "It ain't over til it's over!"

Breaking the Emotional Barriers — This Is Not Your Father's Depression

Many people approach their finances today using philosophies that have been passed down through many generations. For example, the concept of using extra cash to pay off your mortgage early was based on Depression-era economics of the 1930s. In those days, banks were unsafe, Social Security was merely a twinkle in the eye of President Franklin D. Roosevelt, and owning one's home free and clear was the only financial hedge most people could rely on.

Of course, times change. Banks are safe, most people have retirement income and/or Social Security to help out as they grow older and leave the workforce, and a home is merely one of several assets that most people keep in their financial portfolios. Are you living your financial life based upon the philosophies of the 1930s?

Although your elders may have handed down lessons to you, you also need to realize that today's economics are somewhat unique to the last 30 years. Financial structures today are vastly different than they were — even a couple of decades ago — and all different kinds of new tax laws and investment strategies didn't even exist in Grandpa's time.

Evaluating real estate within your net worth

Calculating your net worth can give you a good perspective on your finances. Many people make the mistake of separating their real estate — particularly their homes — from the rest of their assets. If you're like most people, your house is probably your single largest financial asset. For that reason alone, don't ignore it. Chances are, of course, your mortgage is also your single biggest financial liability. Understanding your asset and liability picture is important to gaining the proper perspective for making major financial decisions such as the structure of financing a custom home. Use the following simple method for calculating your net worth:

1. **List all your assets with their values.**

   Make sure you include the current sale value of your existing house, the resale value of your cars, cash on hand, and any stocks, bonds, cash-value insurance policies, or retirement money you may have squirreled away.

2. **Add up the total dollars of all your assets.**

3. **List all your *liabilities* (money you owe).**

   Don't worry about the monthly payment amounts, instead, write down the outstanding balances of your mortgage, car loans, student loans, credit cards, and other debts.

4. **Add up the total dollars of all your liabilities.**

5. **Subtract the liability total from your asset total.**

   This amount is your net worth.

Now that you know your net worth, you can assess how your real estate fits in the picture:

1. **Estimate today's value of all your real estate.**

2. **Add up the amounts of all your mortgages.**

3. **Subtract the mortgage total from the value and this amount is your net equity.**

4. **Take your net equity and divide it by your net worth.**

If the number is greater than .45, then you have too much of your net worth tied up in equity. Even the most conservative investors don't like to keep all their eggs in one basket. Talk with your financial advisor, accountant, and loan officer about the methods for increasing your *liquidity* or available cash and the benefits of doing so.

Acquiring secured debt can be good

Everyone agrees that too much debt can be terrible, but we can think of several good reasons to borrow money if you know you have the ability to pay it back. Unfortunately, debt can easily get out of hand for some people. All debt, however, isn't treated the same.

- ✔ **Unsecured debt** is cash loaned to you strictly on your promise to pay it back with interest (and any other fees that you and the lender mutually agree to). Repayment of the loan isn't guaranteed by any of your property (home, automobiles, boats, and so on). Unsecured loans generally have interest rates significantly higher (the words *loan shark* may come to mind in some cases) than the prevailing prime rates that banks charge their best customers.

- ✔ **Secured debt** is a different animal altogether. Repayment of the cash loaned to you is guaranteed by some item of your property like a car or your home. When debt is secured against an asset, it simply means you're making the asset more liquid. The asset still exists to pay back the debt, and very few banks will allow you to borrow more money than exists in the asset. In other words, you're really borrowing your own money out of the asset.

Because the loan is secured, the lender has less risk and will generally loan you the money at lower interest rates. You can also receive government subsidies for borrowing against assets such as real estate.

Getting on the same page — How banks evaluate risk

Lenders don't think secured debt is bad. In fact, they'd rather bank on someone with secured debt and liquidity (cash in the bank) than someone who has no secured debt at all. Consider this example:

Two neighbors walk into the local bank. Bill Smith has a home worth $400,000. He has paid off his home and has $20,000 in the bank. His net worth is $420,000. Jane Clark, Bill's neighbor, also has a net worth of $420,000, but Jane's net worth is structured differently. Her house is also worth $400,000. But Jane has a mortgage for $300,000, and she has $320,000 in certificates of deposit and mutual funds.

Who gets the bank's attention?

The bank offers Jane its best signature loan for $200,000. She can have the money today with five minutes of paperwork and no costs. Why does Jane get this special treatment? Jane is no risk to the bank; she has ready cash available to pay back her loan if necessary.

Bill, however, is sent down the hall to the mortgage department. He doesn't get a quick loan. Instead he has to apply for a secured mortgage, and then wait for his money while the bank verifies all his information and performs an appraisal on the house. Bill has no means to pay back the loan unless he sells his house, so the bank wants to secure the loan to his house to protect its investment.

So, even though on the surface Bill seems to be the most prudent investor, he's actually a walking, talking risk in the eyes of the bank.

How would you look to your lender? Consider changing your financial profile to look more like Jane's.

Changing perspective — Home equity isn't a savings account

Many people have the erroneous belief that paying off their home is just like putting money in a savings account. Nothing could be farther from the truth. Understanding this concept is important so you can make the right long-term decisions regarding your custom home financing. Your home is *not* a bank!

Why not? Because

- ✔ When you put money into a savings account, you can withdraw the money whenever you want. The only way to draw money out of your house is through some sort of mortgage or credit line, which requires time and often some costs.

- ✔ You can quickly move money in a savings account to better investments as markets change. Equity is stuck in the house and can only be acquired at prevailing rates.

- ✔ You can take out money in savings in any amount — no matter how small — and at your discretion. You can only remove equity in large amounts and at the discretion of the bank willing to loan money to you based upon whether or not you qualify.

Understanding the benefits of liquidity

When it comes right down to it, good old-fashioned fear is the biggest motivator for tying up cash in the home. Some people are afraid they might squander their money, so they increase their payments. Others are afraid of losing their job, so they increase their payments. Many are afraid of the economy changing, so they increase their payments. To counter those fears, consider some basic benefits of keeping your cash liquid instead of burying it into your house:

- ✔ **If you lose your job:** If you lose your job, you'll need to worry about a lot more than just making your house payment. Borrowing extra money on your mortgage for the sake of having cash in the bank can be an excellent form of security. Even though having a house payment that is $500 cheaper may make a financially difficult situation a little easier for you, having a higher monthly payment with $100,000 in the bank will make life a lot easier while you work on finding a new job. You can take your time to assess the situation, and you can afford to take an extra few months to find the right job — instead of feeling like you have to take the first offer to come along.

- ✔ **The economy:** If the economy goes south, then you have cash on hand to bail you out of any situation. If your house decreases in value, so what? You already have the money to use as you see fit. If the interest rates are on the rise, all the better for you because now you can get a higher return on your cash.

- ✔ **Cash squandering:** Getting a professional to help manage small portfolios of money isn't easy. The more money you have to invest, the better caliber professionals will be available to you. They earn a percentage of your portfolio's growth, so they'll work hard to help you save and invest in ways that will benefit you both in the long run.

All this advice assumes that, instead of spending every extra bit of cash like a drunken sailor, you put it aside into a savings account, stocks, a money market fund, certificates of deposit, or other liquid assets. So, instead of running out to buy an expensive new sports car, taking an around-the-world ocean cruise, or gold-plating your plumbing faucets and fixtures, invest your money wisely!

Okay, So You Have All This Cash — Now Manage It

So now you have the cash, and with it comes the added responsibility of managing it. Having wealth can be a pain initially, but have no fear; excellent resources are available. Start with the terrific book, *Investing For Dummies* by

Eric Tyson (Wiley). Then check out some of the many excellent online resources to help you discover how to manage your money. Our favorite is www.motleyfool.com.

Finding and working with a financial advisor

If doing your homework is too time consuming or mind boggling, then don't hesitate to work with a professional financial planner. After you have a decent size portfolio, getting the attention of an experienced professional to help you manage your dough should be fairly easy.

Ask friends and family that are financially secure whom they use for financial planning. Many qualified, experienced professional money managers are out there, but many of them aren't up to the task or don't have your best interests in mind. Do your homework. Ask for referrals and references and check them out. You want someone who will ask you a lot of questions to prescribe the best options. Three basic types of financial advisors are available to you:

- ✓ **Fee planners:** These planners work on flat fees rather than commission-like insurance-based advisors and stockbrokers, and therefore they aren't necessarily tied to selling you any particular kind of investment. Many people believe the lack of commission makes them more objective. Others argue they aren't motivated to earn you the best yields. For more information on fee planners, check out the National Association of Personal Financial Advisors Web site at www.napfa.org.

- ✓ **Insurance-based advisors:** These advisors work for insurance companies or independent insurance brokers, and their primary focus is using insurance policies as investment vehicles. They're often experts in estate planning.

- ✓ **Stockbrokers:** These advisors deal primarily in stocks, bonds, and mutual funds (and they're in business primarily to sell stocks, bonds, and mutual funds). Many larger brokerages have expanded their services to provide other banking and insurance services.

After you find a few financial planner candidates, investigate their credentials and approach. You want to work with someone that takes everything in your financials into account and who is knowledgeable in all kinds of different investments. For example, you'd be surprised how many advisors have little knowledge in real estate.

Stay away from brokers just trying to sell you the hot stock *du jour*. And be aware that there is no way to magically make money. The best advisors advise. Find someone that can spend his or her time educating you so you can make the decisions.

If you can't find one person you trust and you have more than $100,000, spread it around. Try splitting your portfolio among a couple of advisors and see which one performs the best over time. As soon as the trend becomes clear, then shift your assets to the financial planner who does the best.

Diversifying your portfolio

A good financial planner may suggest several strategies, but the primary focus needs to always be *diversification* — splitting your money over many different types of investments. Why? Because no high-performing investment stays that way forever. Diversifying your assets can protect you when a market goes through changes.

Some markets work in the opposite direction. For example, when interest rates rise, the stock market may fall, and vice versa. By investing in a number of different kinds of assets, you can protect yourself from volatility in a particular market and gain a steady return. Studies have shown that well-diversified portfolios have consistently earned 8 to 10 percent during any ten-year period of time in history — from the time of the Great Depression through today, even accounting for the bursting of the stock market Internet bubble. Work with your advisor and discuss the best way to divide your assets, keeping tax ramifications in mind. Different options include

- Commodities (such as oil, orange juice futures, and pork bellies)
- Corporate bonds
- Government/municipal tax-free bonds
- Large-company stocks
- Precious metals (gold, silver, titanium)
- Real Estate Investment Trusts (REITs)
- Small-company stocks

Many advisors recommend investing in mutual funds and annuities because they're a single fund that mixes many categories for you. You can focus these funds toward income or growth. You can also look at an established fund's *prospectus,* a pamphlet explaining all the management details and risks associated with a fund, to see how the fund has performed over time and who is managing the fund.

Exploring alternative investments

Other types of investments can provide good returns outside the standard investment markets. Ask your financial advisor about the risks and benefits of investing in some alternative investments. Check out the following investments that are worth exploring:

- ✔ **Equipment-leasing funds:** Usually available only through a financial advisor, you loan money secured against business equipment, such as computers, furniture, tools, and other assets.

- ✔ **Federal tax credits:** Believe it or not, you can buy someone else's tax benefits and the government guarantees them.

- ✔ **Notes:** Just like a bank or savings and loan, you can loan money to people and get paid interest in return for the risk you take. A *note* is the written promise by a borrower to pay you back with interest. To protect your investment, you can secure notes against real estate holdings.

- ✔ **Limited partnerships:** A limited partnership is a group of two or more people who work in partnership to invest in assets such as large commercial buildings, apartments, and shopping malls. Limited partners, by definition, have only limited say in the operations of the partnership.

- ✔ **Other real estate:** Many people opt to take their money and buy rental properties. See `www.stratfordfinancial.com` for more information on this topic.

You won't find a shortage of investment choices. Not all are financially sound or make sense to every investor. Investigate each one thoroughly and leave yourself with options. Protect yourself by only investing in something you can effectively explain to someone else.

Turning Your House Into a Money-Making Machine

For most people, their house is more than just their home. It represents their single largest investment, as well as their largest source of long-term income. When building a custom home, consider your home an investment. With any investment, you need to make decisions aside from whether or not to invest, such as how to leverage using financing and how to take actions to maximize the return. Many of the questions require research and conversations with professionals like certified public accountants (CPAs), real estate agents, and loan officers. Understanding how a home works as an investment will better prepare you to make the right financial decisions on your custom home to help you maximize your investment return.

More house for less cash — Benefiting from leverage and appreciation

The way people make money on their home is through *appreciation,* which is a return on your investment — the reward for taking a risk with your money. With very few exceptions, you can count on real estate to go up in value over long periods of time. Although appreciation isn't always consistent from year to year, short of some local economic catastrophe, you can typically expect to see the value of your home rise at least 3 to 5 percent per year. Some parts of California and New York have risen more than 20 percent annually in recent years.

You can increase the return on your investment with something called leverage. *Leverage* means using less of your money to make the same profit. You have a choice about how much of your cash is tied up in your house. The dollar amount of the appreciation will be the same regardless of the amount you have invested, but your return can increase substantially using the principle of leverage to your benefit.

For example, consider Jim and Mary, who own a house worth $400,000 free and clear of any mortgages or other encumbrances. Their house appreciates by 5 percent in a year, so they have earned $20,000 or a 5 percent return on their $400,000 investment.

Their neighbors, Tom and Sue, have a house also worth $400,000, but they only have $100,000 equity with a $300,000 loan. They've invested the $300,000 cash that they could have used to pay off their loan elsewhere, such as a diversified stock portfolio. Their house also appreciates the same 5 percent or $20,000 in a year. But instead of the same 5 percent on their investment that Jim and Mary earned, the $20,000 in appreciation represents a 20 percent return on Tom and Sue's $100,000 home equity. Not only is their return better than Jim and Mary's, but also Tom and Sue are free to make other diversified investments with their large chunk of cash.

Protecting your investment by making it marketable

Much of the fun of building a custom home is in creating something that is exactly what you want — it's your dream, after all! If your dream is too unique, it can pose a problem, however, if what you want is unappealing to the rest of the world. A property's value is based upon supply and demand. If many people like your home and want to buy it, then they'll bid against each other and the price will increase. By contrast, if no one wants to own your house besides you, you may not be able to give it away. Keep these major considerations that can impact value and marketability in mind:

- ✔ **Conformity:** If the house is much smaller than the rest of the neighborhood homes, it will be less desirable. At the same time, a home that is bigger than the other homes in the same neighborhood won't attract proportionately more money.

- ✔ **Design:** People want homes that are functional and easy to maintain. A castle with a moat may be great in the French countryside, but will sit on the market forever in suburban Cincinnati.

- ✔ **Location:** Remember the real estate agent's favorite mantra: Location, location, location! Elements such as busy streets or being close to commercial buildings can deter buyers.

Your best financial security is knowing you can sell the house quickly in any market, good or bad. The more the house appeals to a large cross section of people, the more likely you can sell it at a good price in any market. Nearly any house can sell when real estate is in a boom. Ultimately, you want to be able to sell for the highest price possible when the market and the economy are at their worst because that is when you need the money the most.

Understanding taxes — Many parts of a home project are deductible

One of the greatest benefits of home ownership in the United States is the benefit of tax deduction. As long as you intend to move into your custom home when it's finished — using it as your primary residence — you have a number of different items that are tax deductible:

- ✔ *Points* (loan origination fees) on the land and construction loans
- ✔ Interest on the land and construction loan
- ✔ Interest on your permanent loan
- ✔ Property taxes

The Internal Revenue Service (IRS) does limit how much money you can borrow and still deduct in interest and points. The IRS only allows deductions on a loan amount of $1 million or less, but it does allow you to deduct the interest and points on an additional credit line of $100,000 beyond the $1 million.

These deductions translate into real dollar savings for you. Effectively, the government is subsidizing you to borrow money on your house, putting more money in your pocket every month. In states like California, people with substantial income can have tax savings of as much as 40 percent of their monthly payment. Here's a quick formula to calculate your tax savings:

1. **Multiply your loan amount times your interest rate to get your annual interest.**

2. **Multiply your estimated value times your local property tax rate.**

   This amount is 1.25 percent in most states, but can be much more in states like Texas and New Jersey. You can get the information by calling your friendly, local county tax office.

3. **Add the interest and the taxes, and then multiply the total times your combined state and federal tax bracket.**

   You can get this number from your accountant or tax preparer.

4. **Take the amount and divide by 12 to get your monthly tax savings.**

   This tax savings represents real dollars you can use toward your monthly payment.

Sometimes, your deduction can be big enough to reduce your income significantly, moving you into a lower tax bracket and saving you even more money. Ask your CPA or tax preparer to check how close you are to the income thresholds.

If you claim these items as deductions during the build and then sell the home as an investment, you may have to amend your returns and pay back the savings. Check with your CPA or tax preparer about the implications and be sure of your intentions.

You don't have to wait until the end of the year to get the cash from your tax savings. You can increase the deductions on your W-9 so that you're taking home more of your paycheck on a monthly basis. Check with your CPA or tax preparer to figure out your proper withholdings and stop letting the government hold your money for free.

Safely Deferring Financial Decisions Until the End of the Construction Project

Even though planning ahead is crucial to a successful custom home project, you can and need to wait until the project is finished before making certain financial decisions. By deferring some of these choices, you can save yourself from making commitments that may strap you for cash too soon, such as applying a large down payment to your land (see Chapter 3). When you have better information at your fingertips, you can make decisions, such as locking in the rate on the permanent loan or deciding on your permanent loan amount or program. We discuss some of these choices in Chapters 15 and 16.

You have no way of knowing how much your project will actually cost or how long it will take until it's complete. Any estimate before then is strictly an educated guess. The same is true for economic conditions. You can spend a lot of time worrying about where the interest rates are or if your house is holding its value, but you can't do anything to change the situation or predict more accurately. Risk occurs either way, so you have to analyze the situations carefully.

The best approach to dealing with uncertainty is education and careful decision making. Keep emotion out of the decision process. Consult experts and educate yourself about the risks and benefits of each decision. Analyze the repercussions of delaying the decision. Ask these questions when making financial decisions:

- ✔ What is the total amount of dollars at risk on this decision?
- ✔ When will I have all the facts for sure?
- ✔ What is the worst case if I guess wrong today?
- ✔ Is there a way to protect myself against the worst case?
- ✔ What is the cost of that protection?
- ✔ Is it worth it?
- ✔ What course of action will make me sleep best at night?

Chapter 8

Knowledge Is Power: What You Don't Know About Construction Loans Can Hurt You

*T*he nation's big lenders have done a good job of teaching you to shop for a mortgage as if it were a TV dinner that you can pull from your grocery store freezer, only comparing price and forgetting taste. Getting a construction loan, however, isn't quite so simple. In fact, obtaining a construction loan is more like cooking a gourmet meal from scratch for a 20-person dinner party. Although important, the loan's total cost will end up being less important than its functionality. If you ignore details, it could be a disaster.

As you read this chapter, put aside any knowledge you may have of regular home purchase mortgages. Construction loans are different; they have different pricing and structure than loans to purchase existing homes. When you're shopping for a construction loan, relatively few loan officers have extensive experience putting one together, and even fewer firms specialize in these types of loans. Therefore, make sure you're as knowledgeable about the process as you can be before you meet with your loan officer.

Write down every detail (from program information to approval guidelines) and don't be afraid to ask questions several times until you understand everything completely. Your loan officer will probably have to research your answers and get back to you later. Never assume he understands what he is talking about unless he has personally funded at least 50 construction loans over the course of his career.

In this chapter in addition to putting construction loans in the proper perspective, we spell out the differences between construction loan types and take you through all the details. This tour includes picking a lender and understanding the different costs for construction lending.

Exploring Your Construction Loan Options

If you're going to borrow money to build the house from any kind of institution, the money most likely will be in the form of a *construction loan.* This loan replaces your existing land loan. A construction loan differs from a conventional mortgage or the land loans discussed in Chapter 3 in the sense that the bank doesn't give you the money all at once. Instead the bank meters out the money based on the progress of construction. The banks have more risk because taking back a house and selling it if you stop paying them is much more difficult when the home is incomplete.

When dealing with construction lending, don't expect to find much standardization with different lenders' loan programs. Banks pretty much design their own products to fit with their own short-term cash needs. Most banks — small and large — borrow money on giant credit lines and use that money to fund construction loans. Local banks and savings and loans may use portions of their depositor base to fund loans. Because construction loans are by nature short term, banks don't need to sell the loans to third parties (a common practice for long-term mortgage loans). This short-term approach makes standard programs unnecessary in the bank's view.

That said, some similarities do exist among different lenders. All construction loans can fit in the following categories.

Getting it done all in one — Benefiting from a single-close

Large institutional mortgage banks, including IndyMac Bank and Countrywide Financing, created these single-close loans during the last ten years. They figured out that people who build their own custom homes live there longer and default less often than other people. They figured (correctly) that, by offering a great construction loan upfront, they could automatically roll these customers into profitable long-term mortgages. See Chapter 15 for more info on rolling into permanent mortgages.

All-in-One, Construction/Permanent (CP), One-Time-Close (OTC), and Construction-To-Permanent (CTP) — no matter what you call it, the single-close loan has the same features. Some features of the single-close construction loan include

✔ Generally, no re-qualification or re-appraisal is required for these loans. Simply finish the construction, and you get a free permanent loan with no questions asked.

✔ The borrower usually has a variety of popular permanent loan programs from which to choose.

✔ Some lenders offer options for locking interest rates before the house is finished.

✔ The lender may allow the borrower to buy land as part of the process provided she has all the necessary construction documentation ready before funding the loan.

As a result of the ease of the overall process, single-close loans are extremely popular and easy to find in the marketplace. These loans are excellent programs for first-time custom home consumers.

Construction-only loans — The double-close process

In this process — used by most local banks, and the only way to get a construction loan before the single-close program came along — the lending bank provides a short-term loan only for the construction time period. After construction is complete, arranging for another loan for your permanent financing is your job. Unfortunately, having to take this step means going through the process of applying and qualifying for another loan, although your bank may have you prequalify for a permanent loan before it commits to the construction loan.

We *strongly* recommend against this type of loan if you find a single-close option that works for you. Why? Because

✔ This process is potentially more expensive because you have to pay for two loans and all the fees and costs that go with them.

✔ This loan is like a short fuse on a long stick of dynamite. Many things can change in the course of 6 to 18 months while you're building your house. You have no way of knowing whether or not you or the house will still qualify for a permanent loan when the house is finished. A change in the market or interest rates could force you into foreclosure or, at best, payments you can't afford.

✔ The banks that sell this product work on a "buyer beware" ethic and consider permanent loan qualification or getting an unaffordable loan to be your problem at the end.

Our advice? Shop around for an all-in-one deal!

Full documentation versus no-income-qualifier programs

Oddly enough, the best loan programs are available to people who can fully document sufficient income for qualification with tax returns and pay stubs. Documenting your income is always to your advantage, because you can borrow more money at cheaper rates and fees. The key? Be sure you qualify before you apply. (We go into more detail about the qualifying process in Chapter 9.) If you don't pass with flying colors, then a no-income-qualifier program is your next best option.

You may have heard the terms EZ Qualifier, Quick Qual, No Qual, No Income Qualifier, Reduced Doc, No Doc, and the like thrown about. These terms by themselves don't designate actual types of loans. They are merely the lenders' marketing terms to distinguish their own no-income-qualifier programs. And, by the way, these loans aren't necessarily easier or quicker than full-documentation loans. They usually require better credit and take just as long to approve. Here is a breakdown of the specific types of no-income-qualifier loans currently available in the marketplace:

✔ **Stated income, verified asset:** You must state your income and verify liquid assets that meet the lender's requirements. See Chapter 9 for more details on this particular flavor of no-income-qualifier loan.

✔ **Stated income, stated asset:** You can state your income and assets with no further documentation required. Stated loans are usually for people with variable income or who are self-employed.

✔ **No-stated income, verified asset:** You put no income on the application; however, you do provide documentation of your assets.

✔ **No-stated income, no-stated asset:** This loan requires great credit and costs more. You simply state your name and Social Security number.

Poor credit and odd-property options

Consumers with bad credit or properties that fail to meet bank guidelines don't have many options for construction loans, much like the land loans that we discuss in Chapter 3. If your credit is below a credit score of 620 (see

Chapter 9) or if you have recent problems, such as bankruptcy, judgments, or tax liens, you don't have many options in the institutional lending world.

Additionally, any property that doesn't meet with lending guidelines as we outline in Chapters 3 and 9 (for example, property without public electricity) requires a nontraditional lending source. The best option aside from robbing a bank (no, we're not suggesting *that* as an option!) is private or *hard* money. (See "Private money — The last resort," later in this chapter.)

Finding a Good Construction Lender

Picking a loan and finding the right lender can be a bit of a chicken-and-egg process. Do you find the lender and then choose the program, or do you search for the lender that has the program you're looking for? Kevin advises you to explore both directions. If you come across a great loan program, explore it further. On the other hand, if you can find an experienced loan officer with good programs, he may be worth a little more money for the service and education he provides.

Many people make the mistake of looking for someone close by and applying with the first person that makes them feel comfortable. The Internet has also become a new resource for construction financing, but most advertisers don't specialize in construction loans. They're just casting their nets wide for business. In fact, most of the links that come up on Yahoo! or Google are clearinghouses generating leads to be sold to mortgage companies. Read the ads carefully; relatively few companies actually do specialize in construction loans, and you definitely want to find an expert if you can.

Sadly, most of today's lenders are more interested in selling you their loans than helping you make the right decisions. Most loan officers are fast-talking sales guys and gals subscribing to the philosophy that the customer is always right. They'll spend most of their time telling you what you want to hear just to lock you up as a customer.

A loan officer shouldn't be a used-car salesman. He needs to be a professional expert advisor using his knowledge to advocate for your best interest. You wouldn't want your doctor or lawyer lying to you just to make you feel good, would you? Expect the same from your loan officer. Find a loan officer willing to risk your business by telling you hard unpleasant truths that will keep nasty surprises from occurring later in the process. Remember that you have lots of time and money at stake, and you don't want to lose either of those because someone was afraid to tell you the truth or acted upon ignorance. Start with a loan officer that listens to what you're saying. If he is talking about programs before asking you about your situation, then steer clear fast.

Choosing a broker or a bank

A *mortgage broker* is an independent loan originator that helps you, the consumer, and can submit loans to many different lenders. Mortgage brokers act as middlemen and are paid primarily by the lender via wholesale pricing on loan programs, which brings up the age-old question: Do I need to eliminate the middleman? Drum roll please. And the age-old answer: It depends! You may get lucky and qualify with the right bank that has the perfect loan for you in your circumstances. It's not likely, but it could happen. The question is, if you didn't ask around and investigate, how would you know if it was the best program for you? Consider the following reasons why bypassing a mortgage broker and going directly to a bank can be limiting:

- Bank loan officers are salaried (not commissioned) and tend to be the least experienced loan officers in the business.

- Banks are product oriented. They have only their programs and, if you don't fit or don't like them, they'll send you away having wasted your valuable time.

- Banks tend to focus only on the construction loan, ignoring other ways of helping your finances.

- Any documentation you give directly to the bank must be used by the bank — even if your situation changes or can be represented more favorably.

- The bank considers depositors to be their customers. To the bank, you're just another liability. (See Chapter 16.)

Mortgage brokers today fund 65 percent of U.S. mortgages because they provide great benefits to consumers, including the following examples:

- Mortgage brokers have access to almost every loan program in the marketplace, providing a one-stop shop.

- Mortgage brokers only get paid if your loan closes, giving them incentive to get the job done.

- Mortgage brokers are required by law to disclose the fees they make, so you know exactly what they're getting paid.

- Good mortgage brokers are accustomed to comparing and contrasting many different loan programs — finding the best one for your situation, regardless of the lender.

- Mortgage brokers can act as a filter, helping to determine how to present your package in the most favorable way to the bank.

- Mortgage brokers work with you months in advance to help prepare your package.

✔ Good mortgage brokers may identify other loan programs for your situation, such as a refinance to help your cash position.

✔ Many mortgage brokers are dependent upon referral business, making them more concerned with your happiness.

Although mortgage brokers aren't necessarily less expensive than going directly to a bank, banks do offer discounts to brokers in the form of wholesale pricing, so the terms can be close. In construction lending, a broker who knows his way through the process and can get you qualified with a good program is worth far more than a bank that offers a discounted program for which you don't qualify.

Not every mortgage broker is worth your respect. Like any high-commission business, there are plenty of slimeballs. Only a seasoned professional with experience and knowledge in construction financing is going to be an asset in helping you get a good loan for your situation. (More about this topic in the next section, "Testing a loan officer's knowledge.") However, if you can't find anyone who meets these qualifications, first educate yourself, using this book as a guide. Then, armed with this information, pick an honest, willing mortgage broker with access to many lenders. Kevin's book *What the Banks Won't Tell You* (Grady Parsons Publishing) — available at `www.stratford financial.com` — has a great test for nonconstruction loan officers.

Testing a loan officer's knowledge

Regardless of whether you work with a bank or a mortgage broker, you need to be assured that the loan officer is giving you accurate information and she knows what she's talking about. Use this book as your guide for testing your loan officer's construction loan competency. Use these five test questions for doing just that, along with references to the chapters for the answers:

✔ What is the difference between LTV and LTC, and how does it relate to your best programs? (See Chapter 9.)

✔ Why is title insurance more expensive on a construction loan? (See the section "Understanding All the Fees" in this chapter.)

✔ What is the difference between a *voucher system* and a *draw reimbursement system?* (See Chapter 10.)

✔ Calculate a nine-month *interest reserve.* (See Chapter 9.)

✔ Why is an *indemnification agreement* necessary? (See the section "Getting the loan after construction starts" in this chapter.)

If you're fortunate enough to be dealing with someone who understands and can articulate these issues, then you're already ahead of the game. If after you pose these problems, your loan officer is standing there with a glazed look in his eyes, run — don't walk — to a different company immediately.

Most loan officers haven't been extensively involved with construction loans. Regardless, you may find someone that has had some experience. Whether or not this person can answer all these questions exactly isn't so important. What is important is that he doesn't try and slide lies by you in hopes you don't know what he's talking about.

Find someone who tells you she doesn't know all the answers, but she can happily research the issues and get back to you. Honesty and patience are the signs of a trustworthy professional. Make sure she also asks you as many questions as possible. You can best judge your loan officer by the questions she asks rather than by the answers she gives.

Getting value added — Education and experience are worth the money

A small number of mortgage brokers in the country act as construction loan consultants. Kevin's company, Stratford Financial, is one of these companies. Consulting mortgage brokers such as Stratford expertly guide you through the entire process. Their knowledge comes from seeing hundreds of scenarios in construction lending. By understanding patterns, they can almost see the future because they have seen something similar before.

They sit down with you at the very beginning of the process when you're just thinking about a custom home. They can analyze your situation as it sits today and design a program that takes into account every variable and road bump you may encounter along the path until move-in day. They then take this data — combined with your personal financials — and develop a specific plan that assures you the most cost-effective method of succeeding with all the financial steps of your project including risk management, cash flow, and lender approvals.

Sometimes the loan fees with these mortgage brokers can be a little more expensive than going directly to the lender; however, often they can find better programs and structure your project in ways that can save you significant amounts of time and money. Furthermore, you'll have peace of mind knowing you have an expert advocate on your side when embarking into uncharted waters.

If you have already made major mistakes and failed with a lender or two, find one of these consulting mortgage brokerages — construction loan consultants are accustomed to fixing these problems. After a loan runs into problems with a lender, it becomes difficult to determine where the problems are and how to fix them. If you find yourself in this position, contact the lender's wholesale office that turned you down and ask for the most knowledgeable mortgage broker in your region. You can also contact www.customhome experts.com or the National Association of Residential Construction Lenders at www.narcl.org.

Private money — The last resort

If you have realized that private or *hard* money is the only way to get the loan funded, then you still need to find the right private lender for your needs. The good news: These lenders are relatively easy to find, and they usually don't offer a lot of different options. Most local mortgage brokers or bankers can give you a referral to private money lenders. These private lenders are simply investors that like to make moderate risk interest income by making real estate loans that banks won't touch. The lender drives the process and it can be relatively easy. Your main concerns in this kind of loan are price and terms.

Hard money usually costs around 10 percent, plus 5 to 6 percentage points in fees, which is roughly twice the cost of conventional financing. The rest of the fees run the same as any other construction loan as we outline in the section, "Understanding All the Fees," later in this chapter. The terms that differ are

- ✔ The loan's time length
- ✔ How much the private lender will loan
- ✔ How the private lender gives you the money during construction

Some private money lenders are also more conservative about values with appraisals than others. Ask the referring party about its prior experience with the lender. Try to talk to two or three private or hard money lenders before you fill out an application and move forward.

The Loan Process from Start to Finish — When to Do What

Because a construction process involves many people over a long period of time, you need to be as efficient as possible. Starting too early with banks can cost you money and waste your time. Waiting too long can hold up the start of your build. Remember that the actual process is all about proper timing. Use this step-by-step guide to managing the loan process.

Although exploring the construction loan process early is important in order to be prepared, you need to complete several steps before applying to a bank. By this time you should have your lot (see Chapter 3), have your finances in order (see Chapter 7), and have your plans finished and submitted to the building department.

Deciding when to sell your existing house

For everyone who thought you'd have to sell the house you're living in before you can afford to build, we have good news. Most lenders don't calculate the cost of your current residence when evaluating your construction loan qualifications. You will, however, need to provide a letter that you intend to sell when the new house is finished. Some lenders also accept a letter of intent to rent the house. Either way, qualification won't be an issue forcing you to sell the house early. You can wait until the new house is finished, and you don't have to incur the expense of moving twice.

Because banks don't count your existing house payment during the loan application process, you can access any necessary cash you may need for your build (see Chapter 7) without affecting your qualifications. Simply refinance your existing house to the highest loan possible and take a low payment, adjustable-rate mortgage similar to the ones we explain in Chapter 16. Doing so gives you extra money for the build as well as low monthly expenses during the build. You don't need to stay on a long-term fixed payment when you're going to sell the house soon anyway.

In order to access necessary cash from their existing home, some people choose to take out a *Home Equity Line Of Credit (HELOC)* to get the extra cash out of the house. (We discuss HELOCs in Chapter 16.) A HELOC gives you access to the cash in your home; however, a HELOC does increase your current payment to cover the cash you take, whereas a complete refinance can give you the cash and actually lower your monthly payment.

If you're worried about making payments on both your house and your construction loan at the same time, don't fret. Many banks offer or require you to hold an amount of money in the construction loan called an interest reserve for covering the construction loan payments. If you have an interest reserve, you won't have to make payments on two houses at the same time. You can read more about interest reserves later in this chapter in the section "No payments — Taking an interest reserve."

Some construction lenders offer *bridge loans* to cover cash needed from your existing residence for construction. We advise you to stay away from these loans if you can. Bridge loans are more expensive than the refinance or HELOC options, and the bank secures the bridge loan with both properties instead of one. With a bridge loan in place, you have greater difficulty restructuring your financing situation if you underestimate your project and run out of money. You don't need this type of loan because the refinance and HELOC options can provide you more cash at a lesser cost.

Applying on time

Loan documentation is only valid for 90 days from the time it is signed or initiated. This expiration applies specifically to your credit report and the property appraisal. If you apply with a bank and don't fund the loan before these documents expire, you'll be forced to pay for a new credit report as well as an appraisal extension or *recertification of value,* which is only good for another 90 days. Most lenders won't fund the loan until the permits are in place. And, because getting permits is often one of the most challenging tasks as you prepare to build your new home, applying for the permits is the best benchmark to use for your loan application. We discuss the permitting process at great length in Chapter 6.

Kevin's company, Stratford Financial, uses the permitting process to determine the correct timing for the loan application. Kevin's goal is to avoid double work, so he starts the application roughly 60 days before the borrower is ready to break ground. Because most building departments take 30 to 45 days to review working drawings, Stratford instructs clients to start loan paperwork when they turn the working drawings in for final approval.

If your permits are done and you're ready to break ground, you can apply pretty much anytime. The faster you turn in all your paperwork, and the more complete it is, the sooner you'll get funded. The loan process generally takes roughly 40 to 60 days. The slowest parts are the appraisal and underwriting processes that we describe in Chapter 9.

Getting the loan after construction starts

Most lenders fund a loan when you're in mid-build with no problem. The only issues they may have to resolve are with the title insurance companies. Title insurance is necessary for the lender to insure that its loan is clearly secured against the property in case you default on the loan. The lender likes to be the first one to get paid back if it has to take the property back and sell it. The title insurance starts officially the day you and the bank *close escrow* (finish executing all the paperwork and transferring of money on a prescribed date) the loan and the *trust deed* (a legal paper that protects the lender's interests) securing the property is recorded with the county.

When you break ground or do any work on the property, you create what the title companies call *broken priority.* This term means that any contractor or sub who hasn't been paid as agreed can at any time file a mechanic's lien securing their debt against the property. (We talk more about mechanic's liens in Chapter 11.) The problem is that these contractors and subs can file these liens months after the loan is in place, but, because the work in question took place before the loan, the mechanic's lien takes priority and the contractor has rights to the property before the lender. Understandably, a mechanic's lien makes the lender very unhappy and nearly impossible for the title company to insure the lender.

Most states have a remedy for mechanic's liens with something called an *indemnification agreement.* Most title companies allow you to sign an agreement assuring them that you'll pay any and all claims by contractors and subs for work completed prior to the date of recording the loan documents. The title company will want to examine your financials to make sure you can afford to take this step, but it's relatively simple and, more important, it's free!

Most loan officers don't know about indemnification agreements, so you need to educate them and check directly with the title company. Check on this issue with your loan officer and title company before anyone starts work on your property.

Preparing the paperwork

Keeping all your paperwork handy and organized can save you time and hassle when applying for your loan. (Check out the organization tips in Chapter 2.) You need to gather personal and construction-related documents that tell the lender about you, your contractor, and your project. The following checklist describes the personal documents most lenders require before they'll review your loan package. The items are listed in the order that most lenders stack their loan documentation:

- ❑ Application (supplied by lender; lists all your personal and financial information)
- ❑ Consent form (supplied by lender; allows the lender to verify your information)
- ❑ Signed disclosures (supplied by lender; keeps the lender in compliance with the law)
- ❑ Most recent year-to-date pay stub
- ❑ Two years' W-2s
- ❑ Two years' personal tax returns
- ❑ Two years' business tax returns (if self-employed)
- ❑ Year-to-date profit and loss statement (if self-employed)
- ❑ Rental agreements (if you own rental properties)
- ❑ Three months' bank statements on all accounts, both business and personal
- ❑ Most recent statements for retirement accounts
- ❑ Closing statement (HUD-1) for land purchase

The following checklist includes the construction-related documents most lenders require before they'll review your loan package:

❑ Three sets of plans, including working drawings

❑ Completed cost breakdown (supplied by contractor)

❑ Completed description of materials form (supplied by contractor)

❑ Completed builder statement (supplied by lender)

❑ Construction contract (signed by contractor)

❑ Architect information

❑ Insurance agent information

❑ Contractor's liability policy (see Chapter 2)

❑ Workers' compensation policy or waiver (see Chapter 2)

❑ Course of construction policy (see Chapter 2)

❑ Copies of permits or permit applications

❑ Copies of paid receipts/canceled checks for items paid and work completed

Even if the lender doesn't require all these items from you upfront, prepare these items in case the lender asks for them at the last minute.

Locking in an interest rate

When to lock into an interest rate is usually the No. 1 question on most home-builders' lips, but you usually have the least amount of control with it. First of all, not all construction loans have programs allowing you to *lock* the rate (fixing the rate at a certain percentage while you pursue the loan application process). The programs that do allow you to lock the rate generally lock at higher than the market rate you would get on a refinance loan on the same day. Why? Because banks can't predict what the interest rates are going to be 6 to 18 months down the road when you finish the house, anymore than you can. And being banks, they're certainly not going to accept additional risks that they can successfully pass on to you. Some banks offer lock options if you pay more in the interest rate or pay upfront fees. Much of the decision whether or not you can lock the rate depends upon on your qualifying program.

When you lock, you're betting that the market will get worse during your build time. But how can you decide whether or not that will be the case? First, analyze the economy by reading all the financial reports on www.yahoo.com and in the *Wall Street Journal*. If you're still sane, have your loan officer lay out the lock options on the programs for which you qualify. Figure out the dollars side by side and make your best guesstimate. Understand taking any lock before you're approved may be a total waste of time if the bank turns you down or you pick the wrong program.

Determining the length of your construction loan

Most construction loans are offered for a period of 12 months. Some lenders offer shorter periods or longer, but rarely are they offered for less than 6 months or more than 18 months. But if you want something different than the standard 12-month loan term, you're going to pay a different rate. The 6-month rate is usually lower than the 9-month rate, which is lower than the 12-month rate, and so forth.

Be conservative rather than optimistic in estimating your timing. Despite your carefully constructed building plans, you really don't know exactly how long it's going to take to finish building your custom home. You may save a little money upfront with a shorter loan — thinking your house will be done sooner rather than later — but you could have to pay expensive penalties for going over your time limit when schedule delays interfere with your best-laid plans. Choosing a 9-month period over a 12-month period may save you only .125 percent or $625 on a $500,000 loan amount, but the penalties on this same loan for going past the 9-month term could be as much as .5 percent of the loan amount, or $2,500 each month until you finish. Ouch! Not only that, but you also have to pay the penalties in addition to the interest you're already paying, which can get very expensive indeed.

Understanding All the Fees

So you've probably been wondering what all this financing is going to cost. There is nothing like the irony of the loan officer handing you that piece of paper with the big number that makes you feel there should be a brand-new car in your driveway. Who are all these people and why do they need to be paid so much money? In this section we break your loan down, fee by fee, so you can justify where your money is going. Be aware that construction loans have their own, unique quirks.

Your best guide to your fees will be the original Good Faith Estimate (GFE) completed by your loan officer at the time of the original application. If you're lucky, your loan officer will fill it out by hand right there on the spot so you have it in her handwriting with a copy you can go back to later. See Figure 8-1 for a sample GFE. If the government form the lender uses isn't clear to you, ask your loan officer to write the fees on a simple piece of paper. Make sure your loan officer explains every fee clearly and consistently.

S
fs STRATFORD FINANCIAL SERVICES

www.constructionloanexpert.com

Borrower: ___Smith___

GOOD FAITH ESTIMATE

This list gives you an estimate of most of the charges due at the settlement of your loan. The figures are subject to change. They are based on a value or sales price or total build of $ ___870,000___ and the proposed mortgage amount of $ ___650,000___ . For an explanation of these costs, please refer to the "Settlement Costs and You" booklet.

LOAN TERMS *Construction to Permanent, Stated Income, owner-occupied, 30 yr fixed, 12 month build, no prepayment penalty*

Starting Payment $ __4544.89__ est.   P.I.T.I. $ __5640.72__   Interest rate __7.5__ % est.

801. Loan Origination Fee __1.5__ pts. + $ __0__ =	$	9750
1303. Funding Fee	$	350
1306. Loan Document Fee	$	350
813. Tax Service Contract	$	65
812. Life-of-Loan Flood Contract	$	15
815. Courier Fees	$	50
803. Appraisal Fee	$	750
804. Credit Report Fee (personal)	$	50
804. Credit Report Fee (business-for self-employed)	$	
814. Processing Fee	$	450
902. Mortgage Insurance Premium (1st year)	$	
1101. Settlement Fee (escrow or attroney fee)	$	1000
1108. Title Insurance Premium *Includes Endorsements*	$	3000
1201. Recording Fees	$	100
815. Misc. *Administration + Inspections*	$	750
TOTAL ESTIMATED NON-RECURRING CHARGES	$	16680

The following items are not fees, but may either be collected or retained at the close of escrow.

901. Prepaid interest $ _____ per day @ _____ % (30 days)	$	
1303. Interest *Reserves (12 months)*	$	29250
903./1001. Insurance *Course of Construction*	$	2000
107./1004. Property Taxes _____	$	
1002. Mortgage Insurance Premium (2 months)	$	
905. Misc. _____	$	
TOTAL EST. RECURRING CHARGES AND/OR RETENTIONS	$	31250
TOTAL ESTIMATED CHARGES	$	47,930
ESTIMATED CASH REQUIRED TO CLOSE	$	0

Date _____

_____   _____
Borrower   Co-Borrower

Figure 8-1:
Good Faith
Estimate.

Courtesy of Stratford Financial Services

Paying points

There is no shortage of loan people trying to make their point. (It's okay to groan on that one.) *Points* are upfront fees (one point is equal to 1 percent of the loan amount) charged for one of two reasons:

- ✔ To generate cash at the closing to pay loan officers and origination departments
- ✔ To reduce the interest rate by compensating upfront for the interest the banks pay to their investors

Despite what the TV loan hucksters say, points aren't inherently evil or good, they just simply are. Paying more points can be advantageous if the points create a comparable savings. You and your loan officer need to figure out what is the best deal for your situation. Points are usually broken up into eighth fractions. Check out the following list to see how they're represented. This list can help you with discussions with loan officers on points and rates:

- ✔ ⅛ = .125
- ✔ ¼ = .250
- ✔ ⅜ = .375
- ✔ ½ = .500
- ✔ ⅝ = .625
- ✔ ¾ = .750
- ✔ ⅞ = .875
- ✔ 1 = 1.000

Even though points have some flexibility, you won't find many construction loans out there at zero points. Depending upon the terms of the program and how much risk the bank takes, your points should range between 0 and 3 for a conventional lender. Hard money is more expensive. For zero points to happen, the lender has to increase the interest rates in order to pay a *rebate* to the originating broker or loan officer. This scenario may or may not work to your advantage.

To figure out whether you're getting a better deal on points versus rate when comparing two loans, follow these steps:

1. **Calculate the loan amount times the points on each loan.**

2. **Multiply the loan amounts by .6 to represent the amount of loan you'll use during construction.**

3. **Multiply this amount by the offered interest rate.**

4. **Add the points and the interest together.**

5. **The loan with the lower total is the better deal.**

Most knowledgeable mortgage brokers want to make at least 1.5 to 2 points on a construction loan because it's much more work than a purchase or refinance. Loan officers split the fees with their company so they don't always get the lion's share of the points being charged. Some mortgage brokers who provide consulting services or solve difficult problems may charge more.

You can negotiate points, especially with mortgage brokers, but make sure you negotiate upfront. Waiting until the mortgage broker has done all his work and then beating him up on fees at the last minute is unethical, particularly if he is charging what he disclosed to you at the beginning. If he didn't deliver what he promised, then negotiating may be okay. Also, by negotiating in a hardball fashion at the end of the process, the broker may simply decide to not fund your loan, leaving you suddenly high and dry. Remember that a good loan officer is worth his weight in gold throughout the entire project to help work with the lender.

If you're concerned about being overcharged, negotiate upfront the amount of points the mortgage broker will charge, allowing for extra work such as problem solving and exceptions. Most mortgage brokers appreciate this upfront approach and accommodate you. And because mortgage brokers must disclose their fees by law, you can insure they're making good on their part of the deal. Imagine how great it would be if other professionals like attorneys and mechanics were willing to agree on costs upfront. Yes, we know *that's* not going to happen, but we can dream, can't we?

Points are generally tax deductible on your primary and secondary residence. As long as the loan amount is less than $1 million, you may be in for some extra savings. Check with your accountant or tax preparer to make sure you're taking advantage of all the tax benefits available with this project. Don't you think it's about time the government started paying you instead of the other way around?

Escrow and title are more than other loans

Clients who look at the GFE often notice significant differences between the rates charged for escrow and title by various lenders. Being able to identify this difference is actually a great tool for weeding out loan officers who aren't really that informed about construction loans. The Real Estate and Settlement Procedures Act (RESPA) heavily regulates title fees, so they're pretty much the same no matter which company you use. The title fees are more expensive, however, than the same fees for a purchase or refinance loan. Escrow companies and escrow attorneys often charge more for construction loans because of the additional risk they assume for construction loans and the added work that is involved.

The reason for the added expense in the title insurance has to do with mechanic's liens (see Chapter 11). These liens are remedies for contractors and subcontractors to recoup money if you don't pay them, and they can attach (file a legal claim against) the property. The title company's job is to protect the lender from this happening, otherwise, the lender won't loan you money. The title company accomplishes this by way of an *endorsement*. Various endorsements are required throughout the project every time you take money from the bank. The price of these endorsements can range from one hundred to several hundred dollars. So, although title insurance (including endorsements) for a $500,000 refinance may cost only $1,200, title insurance for a construction loan (including endorsements) may cost as much as $3,000.

If you're working with a loan officer who is well versed in construction loans (exactly the kind of person we advise you seek out), accept the title and escrow company she recommends. You may already work with your own attorney or escrow officer, but chances are that he isn't familiar with the construction lending process. If he stumbles or makes a mistake, *you* will suffer. Using someone else's title company on a construction loan can end up adding two weeks to the process.

My goodness . . . so many appraisal fees

Construction loan appraisals are different than appraisals obtained for a purchase loan or a refinance. For construction loan appraisals, the appraiser has to work from the plans and specifications to assess what the house will be worth when it's finished — a feat much more difficult than assessing a house that is already built. And because custom homes often congregate in neighborhoods of other custom homes, comparable properties against which to compare home values are hard to find. Needless to say, these appraisals take more time and cost roughly 25 percent more than conventional appraisals. The price of an appraisal can range from $350 for small houses in a modest price range, to $2,000 for mansions in multimillion-dollar neighborhoods. If you have a large loan amount — say, more than $650,000 — you can double the cost of the appraisal fees because the lender will probably require two appraisals.

In addition, many lenders review the appraisal because it's their most critical piece of information as we describe in Chapter 9. They may just have it *desk-reviewed,* meaning a reviewer checks all the information stated by the appraiser with his or her own research online. This desk-review generally doesn't cost you any money. The lender may, however, request a field review where the lender has another appraiser go out to the property and state an opinion. This extra appraisal may cost you an additional $150.

Don't order your appraisal too early in the process; your appraisal is only good for 90 days. Although an appraisal can be extended with a recertification of value for an additional 90 days, depending upon the lender's guidelines, this extension can cost you another $150 to $250 depending on whether the appraiser needs to add recent comparable properties to bring it up-to-date. And don't order the appraisal yourself. After you select a lender or broker, have him do it. The appraisal has to be in your lender's name anyway, so if you change lenders, you'll need a *retype* at a cost of $150 — assuming that the appraiser gives you the retype at all. Appraisers tend to be loyal to their lenders, so they make up all sorts of excuses as to why they can't release the appraisal to you even though you paid for it. A threat of legal action usually solves this problem, but paying for a new appraisal is much cheaper than litigation.

Insurance costs

Don't forget to consider the cost of insurance in your construction loan; we cover the various kinds of insurance necessary in Chapter 2. Assuming you aren't acting as your own contractor, you'll have one insurance policy to pay for at the close of escrow — the *course of construction* policy. This insurance policy pays for any damage that occurs to your property during the course of the build, and it must be in place before the lender will fund the loan (the lender protects its investment in the property, as well as your own!). Your insurance agent needs to start shopping for the best policy for your situation early in the process. The price of the policy is based upon the loan amount, and roughly runs $1,500 to $3,500.

Figuring all the little stuff

Various parties charge their small fees in the construction loan. These fees are often referred to as *garbage* or *junk fees*. But, junk or not, they can add up, and they deserve your attention. Some of these fees may be negotiable, but most aren't. Don't let that stop you from asking.

So, what do these fees pay for? Most are for items necessary to cover real costs that occur on the loan. Because consumers have become so points- and fee-sensitive, lenders have felt hard pressed to break the loan fees out into every service they have to pay. The following list details most of the small fees that routinely occur in a construction loan:

✔ **Administration and inspection fee:** We extensively discuss the purpose of this fee in Chapter 10. This fee covers the costs of inspections and wire transfers every time you take money during the construction loan. It may be collected along the way rather than upfront, but if taken at the beginning, the fee is usually $700 to $1,000.

✔ **Credit report:** Each lender that is considering loaning you money has to order your credit report from a credit reporting agency such as Experian. These fees have decreased recently, and are usually less than $50 for all three bureau reports.

✔ **Documents:** The lenders subscribe to document software services that prepare the loan documents for signing. Often they pass on the $175 cost to you.

✔ **Flood certification:** The lender needs to know if your property is in a flood zone to make sure it doesn't require special flood insurance. This fee is a cheap one at $15.

✔ **Funding and underwriting:** Lenders actually give these terms a variety of different names; they charge $400 to $750 in junk fees because they can (and they do). The fees generally represent administrative fees within the lender's particular branch. In his 21+ years in the business, Kevin has never successfully argued them away, so don't feel bad if you can't either.

✔ **Messenger fees:** A whole bunch of paper shuffles from the broker to the lender, from the lender to escrow, from escrow back to the lender, and much more. Sometimes documents can be e-mailed, but the lending industry does a good job of keeping messengers and UPS very busy. Figure $50 to $100.

✔ **Processing:** Mortgage brokers charge this fee to cover their administrative costs because most of the points go to loan officer commissions. Many mortgage brokers use them as processor incentives to improve service to the customer. Figure $350 to $500.

✔ **Recording:** A number of documents have to be recorded with the county, including the trust deed securing the property, as well as the construction loan agreement. This costs roughly $100.

✔ **Tax service:** The lender needs to know that you're paying your property taxes on time to ensure that the property doesn't get *attached* (legally taken) and sold at auction by the government. Independent companies provide this service for $65.

✔ **Wire transfer:** The lender generally wires funds to the title company at closing. Expect to pay roughly $25 using the federal wire transfer system.

Letting the Lender Carry Your Burden

In his many years (and more than 800 successful projects) in the lending industry, Kevin has discovered some philosophies for using construction loan offerings that can benefit you in the long run. The following sections explain it more in-depth.

If someone offers you money, take it

Always apply for the highest loan you can qualify for. Ultimately, you don't know what this project will cost until it's finished. As we explain in Chapter 20, nothing is worse than running out of money in the middle of the project. In addition, you may have some last-minute surprises before the loan gets started that require more money. You can usually cover these surprises with a higher loan, but only if you have structured your loan to qualify for that higher amount.

You may be the smartest construction estimator on the planet and think you're being smart saving two points by borrowing $100,000 less than the bank offers. Chances are, something unexpected will occur and saving that $2,000 will cost you $20,000 while your project sits because you ran out of money, or it will cost you an additional $15,000 to get a new construction loan — assuming you can find another lender that will accept your project.

There are benefits to borrowing more money than you think you need. Taking a higher loan can reduce the amount of cash you have to put in the project at the beginning. (See Chapter 9 for more details.) You don't have to use all the money from the loan if you don't need to, and you can roll to a lower loan amount as we explain in Chapter 15. Consider the extra points a tax-deductible, low-cost insurance policy for your new home.

No payments — Taking an interest reserve

Some, but not all, construction lenders offer you the option of having an interest reserve. (We explain how to calculate this reserve in Chapter 9.) An *interest reserve* is a portion of the construction loan funds that are set aside for making payments on the construction loan during the build. Every month, the lender calculates how much money you have requested, multiplying this number by the interest rate and dividing it by 12 to come up with the amount of interest due for the month. If you have an interest reserve, the lender simply adds the amount to your loan balance. If you don't have an interest reserve, you need to make this payment out of pocket.

Some lenders require the reserve, but for those that don't, here are three reasons why the interest reserve is a good choice:

- ✔ **One less thing to worry about.** You'll be managing the project, your job, and your family, and everything will be in transition. An interest reserve gives you one less bill to keep track of and one less check to write. If you miss a payment, it could be disaster for your credit.

- ✔ **It can increase your loan amount.** Lenders only provide a loan for items covered in your budget, which we explain thoroughly in Chapter 9. The interest reserve is a line item that the lender will happily accept.

✔ **It helps your cash flow.** You're going to make payments on the house you live in during the build. Without the interest reserve, you have to make two house payments along with any other money you have to spend during construction.

Interest reserves are only offered as a part of a construction loan. If your lender doesn't require an interest reserve, ask it if an interest reserve is an option. If your lender doesn't require or offer an interest reserve, you can always take the same amount of money from your savings and set it aside to be used for payments.

Chapter 9

Qualifying: It's the Bank's Way or the Highway

*1*f you're paying cash out of your pocket for your new home, then most of the information in this chapter isn't necessary. But, because only a very small number of custom home projects are completely self-funded, you probably need a lender's help to turn your new home into reality. If you're like most borrowers, you'll probably begin with a local bank or national lender. Throughout this process, your bank will remind you of its own version of the Golden Rule: Them that has the gold makes the rules!

Usually, that rule works out just fine, except, most of the time, your lender doesn't tell you what the rules are until after you've broken them, which is why we include this chapter. Although every lender has its own unique take on the rules, a number of elements are common that you need to familiarize yourself with before beginning the loan process.

In this chapter we explain how banks make decisions on construction loans. We give you a look at bank guidelines for approving construction loans, and we provide you with the math formulas banks use to calculate construction budgets. Finally, we address issues that can come up in the bank approval process — and there are more than a few!

Stepping Behind the Desk — How a Construction Lender Views Your Project

When you apply for a construction loan, the bank doesn't consider you to be a customer in spite of the friendly smile on the loan officer's face. What? How can that be? Actually, the bank's customers are the people who give it money to invest. Banks make their profit by loaning money at higher rates than they pay to investors with as little risk as possible. You're a necessary "evil" involved in making that profit. The money comes automatically when you make your payment, so the entire process for giving you a loan is centered around making sure you and the project are a safe investment.

In this section we give you the banks' perspective on risk and their approach to business. We also explain their approach to your intent for the property.

Why some lenders may seem uncaring

Gone are the days when banks made decisions based upon relationships or their personal knowledge of the borrower. Today, government regulators and statistics control most of the policies and procedures in banking. The result is that fewer institutions are loaning their own money for mortgages; Wall Street actually backs most loans in some fashion, even construction loans. The pooling and selling of money to Wall Street — combined with the savings and loan debacle of the late '80s — has created a much more regimented system for *underwriting,* a term that means assessing the overall level of risk on a particular loan.

Although today's electronic underwriting may seem like a less personal approach, it has allowed the underwriters to be a bit more consumer friendly. The underwriting is now less arbitrary. With the computer modeling taking most of the heat for denials or conditional approvals, the underwriter can be more of a consumer advocate focused on getting as many loans that fit through the system. Essentially, the underwriter can play the good cop while the computer plays the bad cop. (Refer to the "Man versus machine" sidebar in this chapter for more information on the underwriting computer system.)

Our recommendation is to not take the underwriting process personally. In one case Kevin was asking his friend — president of a large bank — for an exception on a loan, and the response was "Gee Kev, I would love to. Please give me a reason I can give to the bank regulators when I get audited, other than just doing a favor for a longtime friend." At that point, Kevin understood that today's loan decisions go way beyond the local scene. Fortunately, with aggressive production bonuses in place, most lending employees want your loan to fit their guidelines as much as you do. The more you design your loan to fit the rules, the smoother the process will go.

Man versus machine

For years, underwriting was considered an art and not a science, but today, science and technology are quickly catching up. Since the turn of the century, most loans go through an electronic underwriting system before being passed over to a human being. These risk-assessment systems take into account data from the performance of thousands of loans over past decades. Based upon credit scores and asset and income information, computer programs determine the likelihood of default on your part. The computer program may raise the interest rate accordingly or suggest denial altogether. In Big Brother fashion, human underwriters are rarely allowed to override the computer decision. The human underwriter's job instead is to verify the accuracy of the input information and review documents that are more visual in nature, such as appraisals and tax returns.

Understanding risk assessment

So, what's the risk? You know you're going to make your payments, right? Actually, so does the lender. The number of consumers that default on their home loans is less then 1 percent according to the Mortgage Bankers Association. In most of those cases, the bank takes back the property to recoup its loss. Banks, however, aren't in the real-estate sales business and they don't want to be. Even late payments affect their process of making money on your loan. The underwriting criteria is designed to weed out the people who have a higher probability of paying late or not paying at all.

The other assessment being analyzed on your file is the ability to sell the loan to investors. Most loans are eventually sold to Wall Street investors that have set specific criteria for loans to buy. If these loans go into default, then the original lender may be forced to repurchase the loan from the investor. Repurchasing a bad loan uses cash that the bank could have used to lend to someone else. And if the sale of a recovered property takes a while for the bank, a default can get very expensive indeed.

How banks view your property

Your property is the major security for the loan that the banks guarantee against default. How much the property is worth in the banks' eyes comes down to marketability or the ability to resell your property if the bank has to take the property back in a *foreclosure*. Banks love properties that are vanilla. For example, the more the house is like every other house in the neighborhood, the happier the banks are. Banks love suburbia. A house resembling all others in the neighborhood may of course be the exact opposite of the adobe igloo with the moat and castlelike turrets on the 120-acre estate you have dreamed about since childhood. And there lies the problem for custom home borrowers.

Ultimately, the lender wants the same thing as you. The lender wants to be able to sell the house fast for the highest price in the worst of economic markets. In order for that scenario to happen, the lender requires a property that has the broadest possible interest to potential buyers. The more unique (or odd) the house is, the more the house requires a buyer with unique taste, someone who is relatively rare compared to the vast majority of home buyers. Fewer buyers mean a longer selling time and lower resale. Lenders have years of data that tell them what houses sell fast and what issues can leave a property desperately waiting for a buyer. Ultimately, the security of the money you invest in your property will be better protected by approaching your design and considering the lender's point of view. You can read more about the design process in Chapter 5.

How lenders view contractors

Lenders have loosened their requirements for contractors considerably over the last few years. They used to require full financial review, including tax returns, bank statements, and a first-born child or two. Some lenders still have a rigorous approval process for contractors because of their concerns about a contractor's financial ability to manage the project's funds. However, because most banks have adopted the draw reimbursement system we describe in Chapter 10, banks are at less risk now that money only gets paid after the work is done.

Most banks simply want to see that the contractor isn't a deadbeat and that he has solid experience building houses, indicating that the home will be finished in a workmanlike manner. Banks also like to see that the contractor's license and insurance (liability, workers' compensation, and so forth) are in order. (See our discussion in Chapter 2 for more information on insurance.) Although the bank provides some level of scrutiny for picking a contractor, you'll want to dig deeper as we discuss in Chapter 2.

How lenders view occupancy

Did you ever imagine that your lender would care who was planning to live in your new home? Well, your lender does care — and your answer has a major impact on your loan. If *you*, the borrower, are going to live in the house, then the house is considered to be *owner occupied*. This type of occupancy earns you the very best rates and the most flexible underwriting terms. Why? Because people rarely walk away from the home they live in, so the lender has relatively little risk that you'll default on your loan.

If you're planning to use the home as a vacation house, then it's referred to as a *second home.* The lender looks to make sure that the house is in a likely vacation area or resort town and not next door or across town. This loan may have terms that are slightly less favorable than owner-occupied homes, but second homes add minimal risk statistically, and you can usually find a variety of good loan programs available.

If you intend, however, to sell your home immediately after it is built, then the lender will consider it as a *speculative* or *spec home.* Even if you plan to rent it or use it as an investment, the lender still calls it *nonowner occupied* and treats it as a spec home. Most conventional lenders don't offer consumers construction loans for spec or nonowner-occupied homes because of the added risk of default. These homes require commercial funding that we outline in the next section, "How lenders view spec projects."

If you aren't sure what you're going to do with the house after it's completed, you can finance it intending to owner occupy and defer the decision to sell until later. To meet the terms required by most lenders, you must agree to occupy the property within 30 days of the home's completion. But few loan agreements specify exactly what occupancy means, or even how long the occupancy must be. Lenders do become suspicious, however, if the property is listed for sale before the loan has been converted to permanent financing. Most construction lenders take a dim view of someone who uses this approach more than once.

If the lender suspects that you're building a property for speculation on an owner-occupied program, your loan may be turned down or, at the very least, you may be hit with a *prepayment penalty* (a fee charged if you sell the house or refinance it within a specified period, usually three years). On a $600,000 house, this penalty could amount to as much as $25,000! In addition, if the lender makes you the loan as owner occupied, and you put the house on the market before it is finished — or, worse, default — your lender can require the note to be paid off in full immediately and sue you for fraud, depending upon state laws. Take our word for it; fraud is one path down which you don't want to walk!

How lenders view spec projects

If you're definitely planning to sell your home immediately after it's built and not occupy it yourself, then you need spec financing. Spec financing comes in two forms: well-qualified and unqualified.

Well-qualified people have

- ✔ Excellent credit
- ✔ Income that is three times the amount of their monthly expenses

- ✔ Liquid assets equal to roughly ⅓ the amount they're intending to borrow
- ✔ A track record of building and selling houses successfully

If you meet these criteria, then you can walk into almost any bank and get a spec loan for your house and begin a strong banking relationship. The loan will probably be at an interest rate of roughly prime plus 1 percent, and the lender will charge you 1 percent of the loan amount in loan fees called *points*. (See Chapter 8 for more information on points.) The lender loans you roughly 75 percent of the value of the finished property as long as you have put in a good chunk of cash as a down payment.

For individuals who don't meet these requirements — that is, unqualified borrowers — private money is the likely path, because banks are unlikely to provide spec financing to this category of borrower. (We discuss private or hard-money financing further in Chapter 3.)

Recognizing What a Construction Lender Really Wants to See

Underwriters are looking for reasons to turn a loan down, not for ways to make sure you qualify. They are given exact "deal killer" rules that can't be broken under any circumstance, and they do check first to make sure that the borrower's package meets all these rules. Rules can include such absolute requirements as:

- ✔ Borrowers must have a credit score of 620 or above.
- ✔ The loan-to-value ratio must be 80 percent or below.
- ✔ The property can't be *off the grid* (not connected to a conventional electricity utility).

If the borrower passes this first test, then the underwriters look at guidelines that may be a bit more gray in nature. Guidelines can include such flexible requirements as:

- ✔ The borrower's debt-to-income ratio must be less than 42 percent.
- ✔ The borrower needs to have an average bank balance of, say, $5,000.
- ✔ The borrower needs to have some previous experience in building a custom home.

Unlike the case of rules, underwriters have some discretion where guidelines are concerned. They can examine the file to see if it fits both rules and guidelines in all areas — credit, income, assets, and the construction project — and then make a decision to either approve or deny the loan, based on their best judgment.

With any given lender, you only have one chance to show off your stuff. If your information doesn't fit a lender's rules and guidelines, you can't usually go back and resubmit your loan application with different information unless you can document reasonable and specific reasons for the change. Submitting your package through many different brokers can also cause problems. If two loan applications from two different brokers are submitted with contradictory information, the lender will be forced to turn down both applications.

We recommend that you find a knowledgeable, honest broker who will openly communicate information about your application. She can prescreen your file and suggest ways to meet the lender's guidelines in an ethical manner. Stick with that loan officer and make her earn her money. Do keep in mind, however, that what may be a hard rule for one lender, may be a flexible guideline for another. A good mortgage broker knows these differences and can submit your loan package accordingly, which is even more reason to be very careful when you decide which broker to cast your lot with. We talk more about mortgage brokers in Chapter 8.

On your credit report

Most lenders use credit scoring for underwriting. They use the three major bureaus in the United States that report credit: Experian, TransUnion, and Equifax. These companies all use a computer modeling system to assess your credit and predict how you'll perform on your loan.

The Fair Isaacs Company created the system for Experian, and your credit score is commonly referred to as a FICO score (the other bureaus have their own brand names for credit scores, including Beacon and Empirica).

The underwriter looks at the scores reported by the three bureaus for both you and your spouse. She then determines the middle score of the three for each of you and uses the lower of the two scores. Scores fall into the following categories:

- ✔ **Excellent (720+):** This score qualifies you for just about any program offered today.
- ✔ **Good (680–719):** With a few exceptions, most programs from most lenders can be had with this score, including no-income-qualifier (NIQ) loans that we address in Chapter 8.

- **Fair (620–679):** This score limits your choice of lenders and programs. Many NIQ programs require scores higher than 680 or 640 for full documentation. A couple of lenders do consider scores down to 620 for both, including IndyMac and First Horizon Bank.

- **Subprime (619 or less):** This category is sometimes referred to as B-paper. Almost no institutional lenders write construction loans for people with this credit score.

Credit scores are heavily impacted by three issues in order of importance:

- **Derogatory (bad) credit:** Late mortgage payments within the last 12 months do the most damage, with late payments on car or student loans also doing major damage. Late payments on *revolving debt* (credit cards or store accounts) may not be deal killers, but collection accounts, tax liens, bankruptcies, and foreclosures do affect you if they're recent. If you have these problems, consult with your loan officer before paying off anything. If you try to fix the wrong thing at the wrong time, you can cause your score to drop even more.

- **Credit card balances:** The best scores go to individuals who use credit wisely, and who keep their credit card balances under control. Ideally, you need to have a fair amount of credit available with the balances equaling about 35 percent of the available credit limits. Maxing out your cards can drop your scores 20 to 50 points, so try and avoid those half-yearly Nordstrom sales (or at least be sure to pay back your credit card quickly!).

- **Inquiries:** Whenever you apply for credit — for a car loan, a credit card, or even to rent an apartment — your prospective lender makes an inquiry to check your credit report. This inquiry has an impact on your credit scores. The more inquiries you have, the more you smell like someone in financial trouble to the computers. Keep your new credit inquiries to a minimum. You can shop an unlimited number of mortgage companies within a 30-day period, but you always need to know who is looking at your credit. After you know your scores, you can tell anyone who needs to know without running it again. Beware of Internet credit-checking services because they also generate unwanted inquiries.

Fortunately, even if you have less than stellar credit scores, you can improve them — often, in a matter of days. The following list describes the best ways to repair your credit if you've broken it somewhere along the way:

- **For derogatories:** If your credit report contains some incorrect information, get a letter saying so from the reporting company. Sometimes, you can call a company and plead for a letter; just don't bother pleading through the regular customer service number — they have heard it all before. Call the CEO's office. You won't get the CEO, but you may get a letter from a sympathetic vice president if you paint the right sob story and appeal to his humanity. Getting mad doesn't work, but crying can be very effective (we're not joking here!). Always make sure the letter is on the company letterhead, references the account number, and says the account was paid as agreed.

✔ **For balances:** Only two of the three major credit bureaus change their scores if you pay down the balances, but that's okay because you only need to boost two scores to increase the middle one. If you're in a hurry, make an electronic payment by phone or Internet. Next, call customer service for the creditor that you just sent the payment to and have a representative fax you a letter on company letterhead referencing the account number and stating the new balance.

If these methods don't work for you, or if your issues relate to inquiries, tax liens, bankruptcies, foreclosures, and the like, have no fear. Time is always on your side. The credit bureaus sweep their records every 90 days or so, and *everything* negative drops off your report in seven to ten years. You may think that length of time seems like a long time, but it beats *forever!* More good news: Most issues older than two years — even foreclosures and bankruptcies —have minimum impact on your credit score, especially if you have re-established a solid record of on-time payments and kept the level of your overall debt under control. Talk to your loan officer about ways that you can raise your scores and the necessary time frame.

On your tax returns

Income is important to construction lenders, but less so when your credit is good and the proposed loan amount is a low percentage of the property value. Full-documentation loans still get the best rates and terms, but not by much. No-income qualifier loans (see Chapter 8) provide a simple alternative when full documentation is too cumbersome, or won't support the income necessary for qualification. Not every lender looks at full documentation the same, but at least most lenders look for some standards when calculating your gross monthly income (GMI):

✔ **Salaried:** Lenders look at two years' W-2s and tax returns accompanied by the most recent pay stub. Lenders want stable employment for two years and give full value to the current salary. Bonuses and overtime are averaged over 24 months.

✔ **Commissioned:** People with variable income have to provide their most recent two years' tax returns. Their income is averaged, and business expenses listed on the tax returns are deducted from income. Lenders want to see consistent or increasing income.

✔ **Self-employed:** Anyone who owns 25+ percent of a company is considered self-employed. The lender asks for two years' personal and business tax returns, as well as a year-to-date profit and loss statement (YTD P&L). The lender adds all three totals together, as well as any depreciation. The total is divided by 24 plus the months covered in the P&L and added to personal income.

✔ **Other income:** Nothing gets missed by lenders; they look at any other income or losses you have from rental properties, investments, partnerships, and other sources. Two years' averages are used, and your income will be adjusted by the profits or losses.

Construction lenders want to make sure you have enough income to make timely payments for your new home, as well as any other payments you have to make, such as for cars and student loans. To figure this amount, they use debt-to-income (DTI) ratios. Two ratios are represented, but only the second ratio — the *back-end ratio* — really matters. This ratio tells the bank what percentage of your income will be needed to cover PITI (principal, interest, taxes, and insurance), as well as all your other monthly credit payments. Here are the steps to calculating the back-end ratio:

1. **You need to know the loan payment.**

   Estimate a loan amount (you have to start somewhere) and find out the current interest rates from your loan officer. Have your loan officer calculate the principal and interest payment, or you can use one of the many Internet calculators like www.mortgage-calc.com.

2. **Calculate the property taxes.**

   Take the house's projected future value and multiply it by 1.25 percent. This amount is the annual tax rate in most areas; you may need to adjust it slightly in your own neighborhood. Divide this number by 12 to get the monthly amount.

3. **Calculate the insurance.**

   Take the loan amount and multiply it by 0.35 percent. (Lenders use this number as an estimate.) Divide by 12 for the monthly amount. Add the amounts from the first three steps to get your PITI.

4. **Add in any installment payments that still have ten months or more to go, such as car and student loans, as well as minimum payments on your credit cards.**

   If you have your own business that pays for these items, they don't figure into the ratio.

5. **Take the total and divide by your gross monthly income.**

   If the number is bigger than 0.45, you probably need to consider a no-income qualifier loan of some kind.

This step-by-step version is a simplified, back-of-the-envelope description of debt-to-income assessment, and many more variables can affect it, depending on the complexity of your income and investments. If you're at all unsure of where you stand, meet with an experienced loan officer and walk through the information until you both understand it. By doing so you can avoid surprises when the loan is underwritten.

In your bank accounts

When it comes to your bank accounts, you need cash, cash, and more cash! (You can also take a closer look at Chapter 7 to discover how important cash is.) Banks know that the surest way to protect a construction project from utter disaster is a big pile of readily available money, which means cash in your bank account. Note the following areas of concern related to underwriting for cash:

✔ **Cash in the project:** The bank evaluates all other cash requirements after it accounts for any money needed for the project. The section "Calculating the Loan Amount and Cash," later in this chapter, can help you determine this number. This money needs to be deposited in escrow before the bank will fund the loan.

✔ **Cash-reserve requirements:** Some bank guidelines require very specific cash reserves. *Cash reserves* are the funds you have left in savings or investment after you put all the required funds into the project. The guidelines vary depending upon the loan type and the institution making the loan. Banks commonly base reserves on multiples of the PITI. Most lenders want to see anywhere from 2 to 24 months PITI verified as reserves. The number of months depends on the loan product's risk. No-income qualifier loans and low credit scores can boost the amount of reserves required.

✔ **Money to support income:** The banks want to know that the cash you have in the bank reflects your earnings. This information is especially important when using a loan where you aren't providing any income documentation. The guidelines for this amount vary from bank to bank. Often the bank doesn't have clear guidelines, simply saying that the situation needs to make sense. For example, if you're claiming to make $12,000 every month, the bank may be concerned that you average less than $5,000 in cash in your bank account. Sometimes the bank sets a specific rule, such as having two months' stated income in the bank (in this example, $24,000).

Ask your loan officer to help you determine the specific amount of cash necessary to qualify *before* you provide any asset information such as bank statements. If you show the bank that your assets are insufficient, you may be stuck — no matter what you do to build them up later. Discussing the guidelines with the bank in the abstract is okay, but the bank must use any documentation that you give it. A mortgage broker may have more flexibility in only providing the most recent information to prospective lenders.

When banks verify your *liquid assets* (cash and assets that can be turned into cash quickly, such as stock or money market funds), they want to make sure the money has been there for a period of time and not recently borrowed to beef up your account, causing the need to make payments or pay it back. When new money magically appears, the bank always wants to identify the *source of funds.* The underwriter looks for the 60-day average balance to determine if the money is *seasoned.* If you don't currently have seasoned funds, you may have to wait 60 days before applying for the loan.

An experienced loan officer knows the banks' quirks they're working with and can guide you through the best process to meet their particular cash guidelines. For example, some banks accept unseasoned funds if they came from a line of credit secured by real estate. The earlier you address the issue of cash with your loan officer, the better chance you have of fixing the problem before the lender sees it.

Lenders don't treat all money equally. Lenders give different values to your liquid assets:

- **Checking and savings accounts:** Lenders accept 100 percent of the seasoned balance in these accounts.

- **Stock:** Lenders accept 100 percent of the vested, seasoned balance in these accounts, provided they're publicly traded and verifiable.

- **Retirement and 401(k):** Lenders accept only 50 to 60 percent of these balances, accounting for taxes and penalties assessed by the government on early distributions.

- **Company accounts for self-employed:** Some lenders accept 100 percent of the seasoned balance in these accounts, others only accept 50 percent, and still others don't let you use them at all. You can sometimes move the money out of your business and into your personal account with documentation and a letter from your company's CPA stating it will do no harm to the company.

On the appraisal

The appraisal represents the marketable value of the lender's security — your new home. Custom home appraisers have the difficult job of determining the value of a house that doesn't yet exist. Truthfully, appraisers can't tell you what the real value is for a house, because the only true, real value is what someone is willing to pay for it. What they *can* do is give you an assessment of your home's probable value based upon comparable properties that have sold in the area within the not-too-distant past.

An *appraisal* is a regulated form that describes every detail of your property based upon the appraiser's physical inspection, as well as the plans that you provide. The appraiser probably also asks you for a breakdown of the cost for your build and a description of the materials you're going to use. The bank looks at the appraisal's first page to make sure that the property's planned construction satisfies bank guidelines.

Lenders require the appraiser to find *comparable* properties, that is, properties that are close in size to your own, that are within a mile in location, and that have sold within the last six months. Finding properties that meet these rules isn't always possible, but the appraiser needs to get as close as he can.

Perhaps the most important page of an appraisal is a graph of three or more comparable properties, adjusting their values based upon the ways they are better or worse than your custom home. Look at Figure 9-1 for an example of how these adjustments impact the adjusted value of the subject property.

The appraisal is the most subjective part of the loan process, and the bank often reviews an independent appraisal and decides to use a value that is somewhat less than what is stated. If the bank uses a reduced value to qualify your loan, the bank lowers your loan amount forcing you to make up the difference out of pocket. Never assume that the bank will take the appraisal at face value — the process can sometimes be arbitrary and perhaps even a bit unfair.

Many people pressure appraisers to push their property's value as high as possible, which can backfire, however, when word of these tactics gets back to the lender. If the lender feels the property value is stretched, you can be sure that the value will be drastically cut, lowering the loan amount. Our advice is to let the appraiser be a bit more conservative, if the value is enough to support your loan. It works even better when your loan officer knows which appraisers your lenders favor the most.

UNIFORM RESIDENTIAL APPRAISAL REPORT

Valuation Section

COST APPROACH

ESTIMATED SITE VALUE		= $	500,000
ESTIMATED REPRODUCTION COST-NEW-OF IMPROVEMENTS:			
Dwelling 1,859 Sq. Ft. @$ 200.00	= $		371,800
164 Sq. Ft. @$ 25.00	=		4,100
	=		
Garage/Carport 624 Sq. Ft. @$ 50.00	=		31,200
Total Estimated Cost New	= $		407,100
Less Physical Functional External			
Depreciation	= $		
Depreciated Value of Improvements	= $		407,100
"As-is" Value of Site Improvements	= $		0
INDICATED VALUE BY COST APPROACH	= $		907,100

Comments on Cost Approach (such as, source of cost estimate, site value, square foot calculation and for HUD, VA and FmHA, the estimated remaining economic life of the property): Physical depreciation is calculated by the age Reproduction cost figures obtained from Marshall & Swift Cost Handbook and local builders. Land to value ratio is typical for the area and is due to a lack of similar available building sites. Proposed construction - estimated economic life will be 100+ years when complete.

SALES COMPARISON ANALYSIS

ITEM	SUBJECT	COMPARABLE NO. 1		COMPARABLE NO. 2		COMPARABLE NO. 3	
Address	6841 Donner Drive Any Town, USA	23 Blackbird Court Any Town, USA		7131 Red Hawk Road Any Town, USA		1903 Daisy Drive Any Town, USA	
Proximity to Subject		0.42 miles		0.91 miles		0.11 miles	
Sales Price	$ N/A	$ 1,250,000		$ 990,000		$ 923,000	
Price/Gross Living Area	$ N/A	$ 399.87		$ 501.52		$ 542.94	
Data and/or	NDC Data	MLS/NDC Data/Tax Records		MLS/NDC Data/DataQuick		MLS/DataQuick/NDC Data	
Verification Source	Inspection	DOC #52847		DOC #154494		DOC #98473	
VALUE ADJUSTMENTS	DESCRIPTION	DESCRIPTION	+(−)$ Adjust.	DESCRIPTION	+(−)$ Adjust.	DESCRIPTION	+(−)$ Adjust.
Sales or Financing		Conventional		Conventional		Conventional	
Concessions		Financing		Financing		Financing	
Date of Sale/Time		06/2004		04/2004		05/2004	
Location	Good	Similar		Similar		Similar	
Leasehold/Fee Simple	Fee Simple	Fee Simple		Fee Simple		Fee Simple	
Site	7,319 SF/Slope	7,374 SF/Slope		7,378 SF/Slope		6,726 SF/Slope	
View	Wooded	Wooded		Superior	−100,000	Superior	−100,000
Design and Appeal	Contemporary	Similar		Similar		Similar	
Quality of Construction	Good / Average	Superior / Luxury	−150,000	Good / Average		Good / Average	
Age	2004	2004		2000		1945	+25,000
Condition	New	New		Slightly Inferior	+25,000	Inferior	+50,000
Above Grade	Total Bdrms Baths	Total Bdrms Baths	−25,000	Total Bdrms Baths		Total Bdrms Baths	
Room Count	5 3 2.5	8 4 3.5	−10,000	7 3 2	+5,000	6 3 2	
Gross Living Area	1,859 Sq. Ft.	3,126 Sq. Ft.	−127,000	1,974 Sq. Ft.	−12,000	1,700 Sq. Ft.	+16,000
Basement & Finished	Storage	Similar		Similar		Similar	
Rooms Below Grade	N/A	N/A		None		None	
Functional Utility	Good	Good		Good		Good	
Heating/Cooling	FAU/None	FAU/None		FAU/None		FAU/None	
Energy Efficient Items	Standard	Standard		Standard		Standard	
Garage/Carport	2 Car Garage	2 Car Garage		2 Car Carport	+5,000	2 Car Garage	
Porch, Patio, Deck,	Porch Patio	Similar		Similar		Similar	
Fireplace(s), etc.	1 Fireplace	2 Fireplaces	−2,500	3 Fireplaces	−5,000	1 Fireplace	
Fence, Pool, etc.	Fence/None	Fence/None		Fence/None		Fence/None	
Amenities	None	None		None		None	
Net Adj. (total)		+ ☒ − $	314,500	+ ☒ − $	82,000	+ ☒ − $	9,000
Adjusted Sales Price							
of Comparable		$	935,500	$	908,000	$	914,000

Comments on Sales Comparison (including the subject property's compatibility to the neighborhood, etc.): Comp 1 was selected because of it's similar location and new construction status; however, it is significantly larger in living area and superior in quality of materials and finishes. Comp 2 is similar in GLA, lot size, quality, condition and bedroom count but this property has significantly superior views from both levels. Comp 3 is an older residence but considered relevant because of its proximity to the subject, its similar useable parcel and similar living area but this property has a superior panoramic view. SUMMARY: While all sales have numerous adjustments, they are applied consistently to produce a value range considered appropriate for the subject property. All sales given equal consideration in determining the subject's value.

ITEM	SUBJECT	COMPARABLE NO. 1	COMPARABLE NO. 2	COMPARABLE NO. 3
Date, Price and Data	12/2000	06/2002	06/2000	05/1999
Source, for prior sales	$93,000 (Lot)	$140,000	$703,000	$421,000
within year of appraisal	NDC Data	NDC Data	NDC Data	NDC Data

Analysis of any current agreement of sale, option, or listing of subject property and analysis of any prior sales of subject and comparables within one year of the date of appraisal: Pe owner and available data sources, no sales nor listing activity in the past 36 months.

RECONCILIATION

INDICATED VALUE BY SALES COMPARISON APPROACH	$	910,000
INDICATED VALUE BY INCOME APPROACH (if Applicable) Estimated Market Rent $ N/A /Mo. x Gross Rent Multiplier N/A = $		N/A

This appraisal is made ☐ "as is" ☐ subject to the repairs, alterations, inspections or conditions listed below ☒ subject to completion per plans & specifications.
Conditions of Appraisal: Subject to completion per plans. THIS IS A COMPLETE, SUMMARY REPORT PER USPAP STANDARDS, IS INTENDED FOR USE BY THE LENDER/CLIENT IN A MORTGAGE FINANCE TRANSACTION AND IS NOT INTENDED FOR ANY OTHER USE.
Final Reconciliation: Market actions of buyers and sellers are best analyzed by the Sales Comparison Approach. That approach is given greatest weight in the reconciliation. The Cost Approach provides confirmation of value only. The Income Approach was not developed.

The purpose of this appraisal is to estimate the market value of the real property that is the subject of this report, based on the above conditions and the certification, contingent and limiting conditions, and market value definition that are stated in the attached Freddie Mac Form 439/FNMA Form 1004B (Revised 06/93).
I (WE) ESTIMATE THE MARKET VALUE, AS DEFINED, OF THE REAL PROPERTY THAT IS THE SUBJECT OF THIS REPORT, AS OF 07/15/2004
(WHICH IS THE DATE OF INSPECTION AND THE EFFECTIVE DATE OF THIS REPORT) TO BE $ 910,000

APPRAISER:	SUPERVISORY APPRAISER (ONLY IF REQUIRED):	
Signature	Signature	☐ Did ☐ Did Not
Name Lorin S. George	Name	Inspect Property
Date Report Signed July 18, 2004	Date Report Signed	
State Certification # State	State Certification #	State
Or State License # ALXXXXXX State CA	Or State License #	State

Figure 9-1: A sample appraisal form.

Building the Bank's Construction Budget

This section covers all the math, what Kevin calls the *Construction Mechanics* — the first thing he teaches to loan officers and underwriters. Some of the numbers are reliant upon each other so you may have to initially come up with estimates in order to get the final numbers in place.

The key point to understand when looking at the construction budget through the bank's eyes is that the bank underwriter wants every single dollar for construction covered on the day that you close the loan. The biggest risk to the bank is taking back the property when it's partially complete. Because the property is no longer a piece of land nor a finished house, the bank would have a difficult time selling it and recouping its funds. Making sure the construction budget has plenty of money is in the bank's best interest so the bank can use those funds to finish the project if necessary and sell it.

Each bank constructs its budget differently, so you need to ask the bank how it treats each of the components talked about in this section. If you undercalculate, you may have to bring in cash to start the project or you may simply get turned down altogether. Also, make sure that the loan officer you're dealing with understands the terms clearly. Many loan officers are inexperienced with their own programs (much less anyone else's programs) and the terms necessary to make them work.

Our recommendation is to have budget discussions with your loan officer three or more times until you're sure the answers you're getting have been thoroughly researched and are consistent. These discussions may seem terribly redundant, but you only get one shot at the underwriter and, believe us, in this case it's far better to be safe than sorry.

To start the budget process, you need to establish a *cost breakdown* accounting for all the costs associated with the build and financing. We describe each of the cost breakdown components in this section with a corresponding letter ("factor") so you can easily plug them into the formula at the end (see "Totaling up the cost-to-build").

Figuring the land: Factor A

Different lenders evaluate your land differently based upon the time you have owned it. Most lenders want to see the land *seasoned* for at least one year, for them to give you the benefit of any appreciation. If you have owned your land for less than a year, then you need to use the land's original price in your budget.

If you've owned the land for more than a year, use your loan payoff amount for the budget. Few construction lenders allow the land loan to stay in *first position* (before the secured position of the new construction loan on title), and most require the land to be paid off with the new loan. To calculate the payoff, follow these steps:

1. **Check the most recent statement on your land loan and start with the loan balance.**

2. **Add any prepayment penalties or payoff fees. Call the lender if you're not sure whether other fees are due.**

3. **Add one month's payment because you may have interest due when you close the construction loan.**

Soft or indirect costs: Factor B

Soft (or indirect) *costs* aren't included in the actual physical construction of your home. Generally, these costs include such items as permits, plans, and fees. Soft costs can be a bit confusing because some costs, such as builder overhead, can be considered soft or hard. In these types of situations, you have to make the decision when constructing the budget. Your loan officer can probably offer some guidance. Be aware that sometimes the placement of some items may change based upon the lender's guidelines. Pick a section that makes sense to you and keep it consistent unless the bank asks you to change it. The following is a list of standard soft costs for you to total in this section (You can find many specific descriptions of these costs in Chapters 5 and 6.):

- Architecture fees
- Design-review fees
- Engineering fees
- Permits — city and county
- Plan check fees
- School fees
- Soils and geotechnical reports
- Utility connection fees

Hard costs (board and nail): Factor C

Hard costs represent the actual physical construction work and materials used to build your home, including the cost of materials and labor. Hard costs must reflect the overhead and the contractor's profit. Sometimes these specific costs (overhead and contractor's profits) are listed as soft costs but it's easier for the lender if contractor funds aren't split up. Contractors really aren't anxious to disclose their margins anyway; what's most important for you is the total cost. If the contractor gives you a fair price, then his profit margin isn't your concern.

If you're early in the planning process, then you need to use the dollar-per-square-foot method for calculating hard costs, as we outline in Chapter 2. Doing so gives you a good first shot at a number to work with, but expect to make changes along the way. If you've already obtained bids from your contractor and subs (also see Chapter 2), then you need to have a fairly reasonable estimate of your build after you extract any soft costs.

The contingency: Factor D

The _contingency_ is an amount of money set aside for unanticipated expenses. The fact is, about 99 percent of all projects go over budget — contractors consistently underbid projects to get the job, and consumers are notoriously overoptimistic about how much building their new home is actually going to cost. So, most lenders add a contingency to provide a cushion for cost overruns. Most lenders calculate the contingency by multiplying the hard-cost total by 5 percent, rounding this number to the nearest $500.

Some lenders use higher percentages depending on the other terms of the loan. Be sure to get accurate information from your loan officer regarding the contingency and how it's calculated. Be aware that lenders still add a contingency, even if the contractor has already stated his own contingency as a line item in the bid.

If your budget is running tight, you can discuss the option of removing the contractor's contingency from the estimate. Tell the contractor you'll assign the bank's contingency to him to avoid duplication of money being held.

Calculating the interest reserve: Factor E

The _interest reserve_ is a portion of the loan set aside to make construction loan payments during the build. Some banks require it because they know you have to pay to live somewhere during the build, and they don't want you to be stretched by double payments. The interest reserve is their way of insuring that their loan is kept current. Not every bank requires an interest reserve.

Because a construction loan functions like a giant credit line that you draw upon as you go through the build process, you won't be paying interest on the entire loan amount until the very end. Use this simple formula to calculate a rough interest reserve:

1. **Estimate a loan amount.**

   You can start by adding Factors A, B, and C in this section, and then increasing the total by 10 percent. Subtract any down payment you made on the land. Doing so can put you in the ballpark.

2. **Multiply your estimated loan amount by 60 percent.**

   Doing so accounts for the partial use of money over the loan's term. Because you take money out of the loan as work is completed (see Chapter 10), the bank figures you'll use roughly 60 percent of the loan amount during the construction period on average.

3. **Now multiply by the interest rate.**

   Banks offer different interest rates and, if you know the one you're taking, you can use that one. If you haven't picked a rate yet, you can use prime + 2.5 percent. Doing so gives you a good benchmark. You can find today's prime rate at `www.forecasts.org/prime.htm`.

4. **If your loan is for a term of 12 months, then you're done.**

   If your loan is for a different term, you need to divide by 12 and then multiply by the number of months you'll have the construction loan, be it 6, 9, 18, and so forth. You can round this number to the nearest $500 in the early stage.

Loan closing costs: Factor F

In Chapter 8 we include a complete explanation of fees for construction loans, so we only give the condensed version here. Along with all the lender's and title company's fees, you need to account for the cost of your insurance as outlined in Chapter 2. Costs vary depending upon the terms and size of the loan. You can get most of these estimates from your loan officer and insurance agent. In case you don't have access to them, multiply your estimated loan amount by 4 percent. This amount may seem high, but again, it's better to be safe than sorry.

Totaling up the cost-to-build

Time to put it all together. Add Factors A through F from this section. Your calculation needs to look something like this example:

Factors	Amounts
A: Land cost	$125,000 (includes $50,000 down payment)
B: Soft costs	$ 25,000
C: Hard costs	$360,000
D: Contingency	$ 18,000
E: Interest reserve	$ 20,500
F: Closing costs	$ 21,000
TOTAL	**$569,500**

This amount is your total estimated cost-to-build. Take this number with you to the next section of this chapter.

Calculating the Loan Amount and Cash

Now that you have a pretty good estimate of your project's cost (see the "Totaling up the cost-to-build" section), you can figure out how much cash is needed and what size loan is necessary. This section tells you if your project is going to fly through underwriting or come to a screeching halt.

Lenders use the appraisal and the cost-to-build to establish their maximum loan amount. After you understand the formulas they use, you can work them backwards and forwards to find the combination of cost, value, loan, and cash that works best for your situation. At the very least, these calculations can serve as the basis for an excellent conversation with your loan officer.

Basing the loan on finished value — LTV

Loan-to-value, or LTV, refers to what percent of the appraised value the loan represents. LTV guidelines vary based upon the size of the loan and the income documentation. Most lenders loan between 65 and 90 percent LTV for construction. No-income qualifying loans usually drop the LTV by 5 percent or more. Sometimes lenders allow your loan to exceed these guidelines on an exception basis, but expect to pay additional fees for the privilege. Here is a typical full documentation scale:

Loan Amount	LTV
$0 to 400,000	90%
$401,000 to $650,000	80%

Loan Amount	LTV
$651,000 to $999,000	75%
$1 million to $1.5 million	70%
More than $1.5 million	65%

Using this previous example, you can now figure out what the appraised value of your new home needs to be to get your loan approved. Simply take the cost-to-build of $569,500 and subtract the $50,000 down payment already made on the land. Now take the remaining $519,500 — which is the loan amount — and divide by 80 percent as designated in the chart. In this example, the finished house needs to appraise for at least $649,375 for the loan to be approved. If comparable houses are selling for less, you'll need to adjust your budget, or bring cash to close.

Basing the loan on cost-to-build — LTC

Loan-to-cost, or LTC, refers to what percent of the cost-to-build the loan amount represents. If you have owned the lot for more than one year, most lenders will finance 100 percent of the cost-to-build, provided the loan amount doesn't exceed their other guidelines such as LTV.

When you've owned the property for less than a year, the lender wants to see that you have some sort of down payment invested in your project. Many lenders use the same percentage for LTV and LTC, simply allowing the lower of the two to determine their loan amounts. Using the example in the previous section, if your house appraised at $650,000, it would meet the 80 percent LTV guideline. However, because the cost-to-build is less, the lender using this method would only loan you 80 percent of $569,500, or $455,600. You would then need to make up the $63,900 difference, accounting for the $50,000 down payment already made.

Some lenders recognize that you can often build for less than the value of existing houses in the neighborhood. These lenders are usually the national lenders and rarely the local banks. They figure there must be some profit margin in new construction, otherwise, why would spec builders build. These lenders offer an LTC percentage higher than the LTV. This percentage typically ranges from 85 to 95 percent of the cost-to-build. Using the higher LTC, the lender loans at LTC up to the LTV. In other words, regardless of LTC, they don't exceed the LTV guidelines without an exception.

Looking at our example a little closer, if the lender offers 85 percent LTC up to 80 percent of value, you'd first calculate the LTC: 85 percent of $569,500 is $484,075. With an appraisal of $650,000, the LTV allows a maximum of $520,000. Because $484,075 is the lower of the two, the lender would use that as the maximum loan amount. The cash to close would be $569,500, minus the loan of $484,075, minus the $50,000 down payment — requiring the difference of $35,425 to be deposited at loan closing.

LTC makes an excellent litmus test for the quality of your loan officer. If your loan officer can't explain the difference between LTV and LTC, run — don't walk — to someone fluent in this process. LTC ignorance causes more than 50 percent of loans to get into trouble in underwriting. Unfortunately, no formal training classes on the LTC subject are available, so most loan officers have to figure it out on their own from experience. Don't settle for second best when entrusting your life savings with someone you may have just met last week.

Calculating the cash needed for the project

After you have the loan amount, figuring the necessary cash is relatively simple. Take the cost-to-build and subtract the maximum loan amount. Whatever is left is the cash required for the loan. If you have already put that much in the project, then you don't need to do anything else. However, you need to deposit any shortage with the title company before the lender will fund the construction loan.

One of the great things about construction loans is that time works in your favor. You can plan a project on a timeline that allows you to put the cash in a little at a time as shown in Figure 9-2. Some money goes in for the down payment, and you can pay soft costs along the way until the lender is ready to fund your loan.

Lenders generally treat paid receipts for soft and hard costs the same as cash. They happily adjust the budget at escrow to account for any items that may have been paid since the loan process began. Keep in mind, however, that you'll likely need to show real proof of payment. Canceled checks are best accompanied by receipts marked "Paid" by your vendors. Just showing your check register or copies of checks not yet canceled usually doesn't fly with an underwriter.

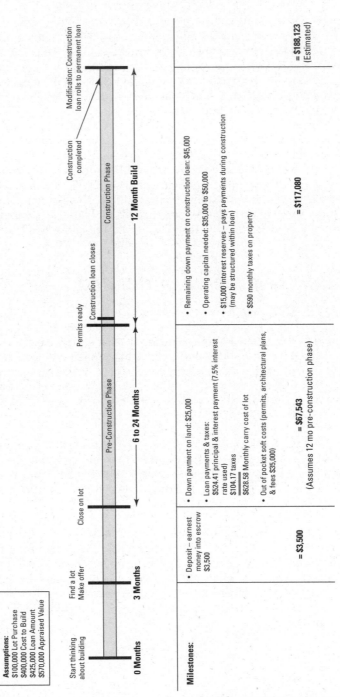

Figure 9-2:
Time-money
timeline.

Assumptions:
$100,000 Lot Purchase
$400,000 Cost to Build
$425,000 Loan Amount
$570,000 Appraised Value

Start thinking about building — 0 Months

Find a lot Make offer — 3 Months

Close on lot

Permits ready — Construction loan closes

Construction completed — Modification: Construction loan rolls to permanent loan

Pre-Construction Phase — 6 to 24 Months

Construction Phase — 12 Month Build

Milestones:

- Deposit – earnest money into escrow $3,500

  = $3,500

- Down payment on land: $25,000
- Loan payments & taxes:
  $524.41 principal & interest payment (7.5% interest rate used)
  $104.17 taxes
  $628.58 Monthly carry cost of lot
- Out of pocket soft costs (permits, architectural plans, & fees $35,000)

  = $67,543
  (Assumes 12 mo pre-construction phase)

- Remaining down payment on construction loan: $45,000
- Operating capital needed: $35,000 to $50,000
- $15,000 interest reserves – pays payments during construction (may be structured within loan)
- $590 monthly taxes on property

  = $117,080

= $188,123
(Estimated)

Solving Other Budget Problems

Budgeting to the lenders' guidelines can be difficult, especially because they vary from lender to lender and change quite often with market conditions. This section contains some challenging issues that can result in major headaches for you along the way.

Landscaping and finish work can kill the project

Landscaping (trees, shrubs, and pools) and *finish work* (crown moldings and built-in home theater systems) generally take the most time and money to complete. Ironically, they usually bring the least amount of value to the appraisal. This poses two problems for a construction loan:

- **Time:** As discussed in Chapter 8, a construction loan establishes a set time frame for the house to be completed before the lender charges penalties (which can be quite expensive, by the way). You can finish most of the house, but if the finish work (described in Chapter 14) or the landscaping (described in Chapter 17) drag out the project, it can get very pricey.

- **Value:** Approximately $50,000 in extra finish work probably adds only $20,000 to your property's appraised value. The same is true for land-scaping. A lender loaning 80 percent LTV only gives you $16,000 toward the $50,000, which means you have to make up the difference of $34,000 out of pocket.

If your budget is running tight, we highly recommend initially keeping your finish materials to a minimum, and they need to be of similar quality to other houses in the neighborhood. You can always upgrade later in the build process if you have the cash or come in under budget. If you're committed to spending money for extra finish, it's never a question of *what* you spend, but rather, *when* you have to spend it.

You can eliminate landscaping from the budget entirely. Most lenders don't require it for completion. Pools, spas, and gardens don't give you dollar-for-dollar value anyway, and they can drag out your project for months. Pay for these expenses out of pocket when the house and the construction loan is done.

Costing — What if I can build on the cheap?

Because banks are concerned about whether they'll be able to recover their investment in you if they have to take back the property in the event of foreclosure, they're very wary of underestimating the build. Many banks have contractors on staff who have the job of assessing your cost breakdown to make sure it seems realistic. If your bank is unsure, it may insist on a higher loan amount or more cash if there is no room for a bigger loan. Your bank may also ask to see the bids from the subs. Overestimating is in your best interest, because running out of money midway through your project can be a disaster.

What if I need more cash than is in my budget?

Most banks have no problem when you overestimate the budget as long as the loan still meets LTV and LTC guidelines. Overestimating allows you to fluff up the cost breakdown to give you extra funds along the way. Because the bank reimburses based upon your estimates (as explained in Chapter 10), you can use this extra money for other expenses, such as land reimbursement or upgrades on finish materials.

The bank generally lets you borrow the entire loan amount even if you don't use it all for construction, so overestimation is another way of funding your landscaping and finish work without taking the risks.

Chapter 10

Show Me the Money

After your project gets going, you somehow have to pay all the people who are building your home and supplying the materials. Because this project likely represents more money than you have ever spent on anything in your life (unless you paid a *lot* of money for your car!), you need to understand how the money moves from one place to another.

In this chapter we discuss how to manage a self-funded project, and we walk you through the different methods that banks use to keep their fingers in your custom home construction pie. We let you in on how the bank likes to see a project proceed. We also share our exclusive way to pay yourself with a vacation when the project is finished. Along the way, we provide you with tips on how to best protect yourself from problems.

Managing a Self-Funded Project

Many people decide to bank their entire project out of their own cash or savings. In fact, 20 percent of all custom homeowners pay for their project without the use of a bank loan. This fact may work when you have enough cash in your investments to pay for it (we're assuming you sold off that dot-com stock *before* the bubble burst!), or you may have sold or refinanced your home to obtain the cash. You can find a discussion of the pros and cons to building a custom home using your own cash in Chapter 7.

The greatest benefit of using your own cash stems from being able to control all the funds yourself. The bad news is that you have to develop your own system of management and accounting for all the many checks that you have to write (and, believe us, there will be *many!*). (We provide suggestions for organizing your accounting systems in Chapter 2.)

Many of the systems already used by banks can meet your needs for keeping financial control over your project. Simply adapt the systems we talk about in this chapter to fit the needs of you and your contractor. And, while you're at it, make sure that you hold the contractor to that system. Not surprisingly, with fewer outside controls, self-funded projects often go over budget or get into trouble. See Chapter 11 for information on working with your contractor and dealing with ongoing budget issues.

Because you don't need to use your cash all at once (banks figure you will use roughly 60 percent of the total budget over the length of the build), over the course of the project, you can take chunks of the funds and diversify them among various interest-paying investments. This money management allows you to earn additional interest on your money while you build your house. For more information on how to manage this cash safely, be sure to take a look at *Investing For Dummies,* 3rd edition, by Eric Tyson.

Working with a Bank

Because most of you will likely borrow money for this project, you'll have to manage the money according to a bank's rules — as the Golden Rule of Credit says: He who owns the gold makes the rules! Construction lenders (which almost always are banks) have designed these rules and systems after many years of trial and error. At times, their rules may seem arbitrary and ridiculous. But after solving many issues with the way banks hand out money (called *disbursement,* by the way), we have figured out that all their rules are designed primarily to protect their investment. Fortunately, the money they have invested in your custom home project is mixed with all the rest of your investment in your custom home, which means that you and the bank have the same interests of a successful project at heart.

Of course, you don't get all the money upfront

Banks only fund money for construction through a *construction loan.* Construction loans are different from other types of mortgages because they're designed to pay money in a metered fashion for the construction of the home. The loan amount is decided based upon the value of the completed project instead of the worth of the home today, like other mortgages. For a detailed explanation of what makes construction loans unique and how they work, read Chapter 8.

Banks structure a construction loan like a credit line. Effectively, you get a credit card with a large available balance that you can use to withdraw money as you need (and, like a credit card, you eventually have to pay it all back, plus interest!). For the most part, you can decide when and if you want to draw on that credit card. The only exception (and it's a *big* exception) is that you do have to meet certain criteria in order to take the bank's money and put it to use on your project. For some reason, banks don't particularly like the idea of putting several hundred thousand dollars into someone's hands without making sure that person's using the money to pay the proper people to do the proper job. You'll find yourself operating under a process of checks and balances, regardless of the reimbursement system your bank uses.

Chapter 8 covers how to develop a construction budget and cost breakdown for your project. That budget and cost breakdown now serve as the basis for how you can draw money from the bank. At this point, the bank has established a loan-in-process account (LIP). Have no fear; every item that you accounted for in your budget is included in the LIP. Now you just have to determine the method and timing for distributing the funds. With the exceptions of deposits for materials, money generally comes from the bank after workers have done their job. The bank requires some proof of payment or completion in the form of inspections and maybe receipts before it hands over (or disburses) the funds.

Preparing for inspection

So, how does the bank know if you're doing what you're supposed to do — especially if your home-to-be is in Ohio, and your bank is in Arizona? The bank performs this little miracle with the help of field inspectors. Field inspectors — who may work directly for the bank, or who may be third parties hired by the bank — visit your job site and physically inspect your progress. Here's a bit of insider knowledge: These inspectors don't always have extensive experience in construction, and most aren't contractors. In fact, some banks contract with the original loan appraiser for the inspections. The point is that *you* may know more about your project than your inspector does.

When you make a request to your bank for money, an inspector will make an appointment to come to the job site. It's our experience that you have the best success with an inspector when you, the contractor, and the inspector walk the property together. Many decisions about the status of project completion may not be black and white. An inspector left to her own opinions may jump to conclusions that may work against you. It's your job to ensure that the inspector sees the project through your eyes.

Even though the bank controls the money, it isn't responsible for the quality of the work. The bank's inspectors are only interested in checking to see that the payments made by the bank are in sync with the project's progress.

Considering that these inspectors only get paid $25 to $50 for these visits, they usually don't put much, if any, time or energy into checking the work quality, and don't expect them to. Furthermore, the bank doesn't really want the added hassle of liability for inspecting the quality of work in case you and the contractor end up in a dispute down the road.

Battling the bank

Anytime people are working through a complicated process with money on the line, disagreements are bound to happen. Most people in the construction lending industry prefer to put their head in the sand and sell you on the ease of their particular process. Don't believe it. In this section we lay out the honest truth about dealing with banks when you find yourself in conflict, along with our insights into potential problems and ways of resolving them.

During the building process, any discrepancies between the original budget and the cost breakdown later submitted to the bank can become a point of discussion with your contractor, as well as your bank. Even though you're building your house and heading your project, the money that you borrow ultimately belongs to the bank. The bank loans it to you in a secured manner and diligently monitors your project to protect its interests. Sometimes the bank agrees with how you see it, and sometimes the bank takes a different view. But keep in mind that you have to see everything the bank's way. The bank is under no obligation to loan you money if it chooses not to, and if everything doesn't go the way the bank wants it to, the bank can cancel the loan in a heartbeat.

Despite the frustration you may feel dealing with the bank's bureaucracy, rest assured that you and the bank have the same goal: completing the house in a timely manner. Use this shared aim as common ground when discussing your difficulties. Remember, although the bank is contractually obligated to provide you with all the money it promised if you meet the commitments you made in the construction lending agreement and budget (discussed in detail in Chapter 8), getting to that point may require flexibility.

The surest way to protect yourself and your project is to have access to plenty of cash beyond your construction loan (from savings, relatives, loans, or the cookie jar in your kitchen) to keep the project moving. Having access to this extra funding allows you to keep paying people and to keep building, putting off the handling of disputes until later, when the pressure isn't quite so intense.

Understanding How the Voucher System Works

The *voucher system* was the primary way of paying construction loans for decades. Many banks, however, have found that the voucher system requires too much paperwork and people chasing. As a result, many banks are shifting to the draw system, which we discuss later in the "Taking a Closer Look at the Draw Reimbursement System" section.

The draw system is a little more lenient than the voucher system, so banks looking to keep tight control on their funds during the custom home–building process — and to make sure that all parties have met their obligations along the way — still tend to use the voucher system.

The voucher system starts with your *cost breakdown,* which breaks down the various stages of your project into specific costs for labor and materials (see Chapter 8 for more details on putting together your cost breakdown). The *voucher* is simply a piece of paper that looks like an oversized check. The front side of the voucher has a place for the date, the amount that the bank will pay, to whom it will pay the amount, and a description of exactly what work or materials the voucher should cover.

On the back of the voucher, you can find a place for the supplier or subcontractor to enter pertinent business information, such as its address and Employer Identification Number (the issuing bank needs this number because it has to report all money it has paid for tax purposes). Last, but not least, you can see a paragraph releasing any rights that the supplier or sub may have to put a lien on the property.

In short, the bank is saying "we paid you your money, now leave us alone." After you have given the voucher to your suppliers or subs, they must fill out the necessary information and provide the original invoices and bids to the bank, when requested. After the bank inspector has verified that the work has been completed, the bank pays the amount requested on the voucher directly to the suppliers or subs by cashier's check.

Most people in the custom home industry agree that the rigid voucher system provides the most accountability; however many banks and contractors find it cumbersome in the increasingly fast-paced and flexible building world. In our experience, the voucher system can be a useful way to manage the payment of construction funds to your project and includes the following benefits:

- ✔ **You can use it when a contractor isn't financially strong.** For example, he may be cash poor or have problems with credit or is in a disputed situation. The voucher system gives you added controls to make sure the money goes where it needs to go.

> ✔ **This system can also help protect you and the bank against *mechanic's liens*.** A mechanic's lien is a legal method for contractors in dispute to guarantee they will get paid. You can find a detailed discussion of mechanic's liens in Chapter 11.

> ✔ **It provides tight budget control handled by a third party (the bank).** This keeps arguments over money to a minimum between you and your contractor because everything must meet bank approval.

But the voucher system isn't always a bed of roses. Here are some key drawbacks:

> ✔ **It requires multiple signatures from multiple parties.** Having to track down all those John Hancocks can be time-consuming or impossible if the parties travel a lot.

> ✔ **It requires strict adherence to the budget, without much room for cost overruns.** This system creates a problem if the project has wide variations from the budget.

> ✔ **It requires the collection of additional paperwork.** Talk about a hassle!

Ultimately, the bank dictates the system it prefers to use for its purposes. The bank normally tries to have one system for efficiency, but some banks do offer you a choice between the voucher system and the draw system, which we discuss in the next section.

Taking a Closer Look at the Draw Reimbursement System

Banks call the most common disbursement process used today *draw reimbursement*. This system puts the bulk of accounting responsibility on you and your contractor. This system has the bank use your original project line-item cost breakdown as a guide and allows you to request reimbursement on a designated schedule after the work is completed.

Some banks use specific schedules of four to seven *draws* (or money payouts) based on the completion of certain milestones, such as foundation or framing. Others allow you to take draws on a monthly basis, collecting partial payment for work and material items that haven't been fully completed.

Unlike the voucher system, banks using a draw reimbursement system generally wire the funds from an LIP account to any account that the borrower wants (try to avoid, however, having the money sent to a numbered Swiss account — that wouldn't do much to instill the confidence of your loan officer!).

The draw system is popular because

> ✔ **It requires less work.** Although this system still needs bank inspectors just like the voucher system, with the draw system the bank doesn't require subs and suppliers to sign off at every disbursement.
>
> The bank checks with the title company to insure that the property is free and clear from any liens at the time of disbursement. The bank does this check, using a 122 Endorsement (which we discuss in Chapter 8). Eventually, you must show the bank all lien sign-offs by the end of the project.
>
> ✔ **It's easier to use than the voucher system.** For example, today some banks let you request draws electronically through the Internet, offering speedier payment and greater convenience.

Banks like the draw reimbursement system because their end is simpler, and it helps keep the project moving. However, the draw system still does have a couple of drawbacks:

> ✔ **Banks scrutinize more.** Because banks take on the bigger risk of everyone getting paid properly, they tend to look a little closer at issues of your contractor's credit and your own available cash.
>
> ✔ **You may have to jump through a few more hoops to get approved.** Because the draw reimbursement system only pays for work completed, you or the contractor may have to lay out money first before the bank pays you back. You may have to show the bank you have more cash on hand before you get approved. (To look at specific liquidity requirements of most banks, check out Chapter 9.)

In the sections that follow, we consider some specific issues related to the draw reimbursement system.

Figuring which costs are which

In order to get your funds reimbursed from the bank, you have to figure out how to speak a bit of the bank's language. The bank will refer to your line item cost breakdown every time you request money. It's referred to as a line item cost breakdown because each section of work and materials (such as rough framing or permits) is called a *line item*. Your bank refers to these line items as either hard or soft costs. Banks agree mostly on what is a hard cost and what is a soft cost, however, we have found that different banks treat certain line items (such as builder overhead) differently. If you have concerns, you can usually contact the bank's disbursement officer and she can explain it to you. Being clear on the definition between hard and soft costs is important because the banks disburse hard and soft costs in different ways.

Soft-cost reimbursement

Soft costs (also known as *offsite costs*) are generally the fastest funds to get paid back from the bank, but they also require the most paperwork. You'll have paid many of the soft costs before the construction loan is in place. These costs — which can include architecture, permits, engineering, and school fees — aren't related to the labor or materials in your project. Because these line items don't lend themselves to inspection, the bank wants specific accounting and copies of paid receipts. Stricter banks want to see copies of cancelled checks for these costs. Most banks reimburse these costs within 24 to 48 hours because they don't have to send someone to the property before payment.

Banks only pay the amount of the proven receipts. If you estimated high, your bank keeps the money in the LIP account (described in the "Of course, you don't get all the money upfront" section, earlier in this chapter). If you underestimated, you have to cover any overages. You can also draw from the contingency we discuss in the following section.

Hard-cost reimbursement

Now, you've officially arrived at the meat and potatoes of your building funds — hard-cost reimbursement. *Hard costs* include the labor and materials associated with the foundation, framing, finish, and complete building of your project. After the build has begun, you can draw upon these funds on an item-by-item basis as you complete each appropriate section.

Each time you make a request, you have to submit a form (see Figure 10-1) telling the bank how much of each *line item* (a particular construction task or job) you have completed. Notice that the cost breakdown in Figure 10-1 reflects the money paid, as well as the money requested. This breakdown helps you and the bank keep track of the percentage of total available funds left to draw on each line item. Before the bank releases the funds, the bank inspector comes out to the property to see if he agrees with your assessment. If so, the inspector notifies the bank, and the bank wires the funds for the amount that you requested to your account. If not, then you and the bank need to discuss the situation until you agree upon an amount.

When requesting reimbursement for line items that can be either hard or soft costs — such as builder overhead or profit — the bank generally pays these amounts out in a percentage proportionate to how far along you are in the entire project. For example, if you have drawn 35 percent of all the hard costs, the bank allows you to draw up to 35 percent of the builder's overhead.

To save money on your project if you're short on cash when starting to build, some banks may advance you up to 5 percent of the hard costs to get the project started. In addition, if your construction loan was put in place after you've already begun to build, you may be able to request an immediate draw to catch up your cash. Discuss these needs with your lender when applying for your construction loan. You can find more information in Chapter 8 about funding after you start your build.

STRATFORD FINANCIAL SERVICES
800-727-6050
www.stratfordfinancial.com

Draw Request Form

Project Type: ☐ Conventional Built Home   ☐ Packaged/Kit Home   ☐ Manufactured Home

Borrower: Tel #   Loan #

Project Address: Architect: Tel#

Contractor/Manufacturer: Tel #   Engineer: Tel#

Line Item Description		Project Costs	Prepaids	Remaining Construction Funds	Amount Requested	Inspector Only % Complete
PRE-CONSTRUCTION 'SOFT' COSTS						
Architect, Engineering, Survey, Energy & Geo.-Tech Fees		$ 35,000	$ 25,000	$ -		100%
Design Review/Plan Check Fees		$ 6,000	$ 5,000	$ -		100%
Permits - City/County		$ 8,000		$ -		100%
Utility Hook-up Fees		$ 12,000		$ -		100%
School Taxes		$ 5,500		$ -		100%
Project Supervision & Overhead		$ 5,400		$ -		100%
PRE-CONSTRUCTION COSTS - PERMIT ISSUED	**Subtotal**	$ 71,900	$ 30,000	$ -	$ -	$ 71,900
DIRECT SITE 'HARD' COSTS:						
Site Inspections - Geo-tech, Structural, Civil		$ 2,000		$ 2,000	$ 2,000	
Temporary Facilities/Equipment Rental		$ 1,000		$ 1,000	$ 1,000	
Project Supervision & Overhead		$ 2,000		$ 2,000	$ 500	
Demolition				$ -		
Rough Grading, Trenching, Backfill*		$ 12,000		$ 12,000	$ 12,000	
Retaining Walls, Waterproofing, Drainage*		$ 10,000		$ 10,000	$ 5,000	
Plumbing - Ground*		$ 8,000		$ 8,000	$ 2,000	
Foundation Poured*		$ 30,000		$ 30,000	$ 7,800	
Concrete Slab Poured*				$ -		
Plumbing Gas Piping				$ -		
FOUNDATION COMPLETE	**Subtotal**	$ 65,000		$ 65,000	$ 30,300	$ -
Underground Utilities* (Sewer/Septic)		$ 3,000		$ 3,000		
Structural Masonry*				$ -		
Rough Framing Materials incl. Rough Hardware		$ 4,000		$ 4,000		
Structural Steel*		$ 25,000		$ 25,000		
Trusses/Components*				$ -		
Plumbing - Top Out including Tubs, Shower Pan*		$ 10,000		$ 10,000		
Fire Protection System*		$ 500		$ 500		
Heating/Ventilation/AC - Rough*		$ 15,000		$ 15,000		
Electrical - Rough*		$ 12,000		$ 12,000		
Fireplace*		$ 1,500		$ 1,500		
Material Package (Kit Home) Delivered**				$ -		
Exterior Stairs/Decks*		$ 2,000		$ 2,000		
Rough Framing or Kit Assembly Labor**		$ 30,000		$ 30,000		
Plumbing Installation Labor				$ -		
ROUGH STRUCTURE COMPLETE	**Subtotal**	$ 103,000	$ -	$ 103,000	$ -	$ -
Waterproofing				$ -		
Roofing - Built-up/Tile/Shingle - Specify:		$ 14,500		$ 14,500		
Caulking		$ 1,500		$ 1,500		
Windows & Exterior Doors*		$ 20,000		$ 20,000		
Garage Doors		$ 2,000		$ 2,000		
Skylights				$ -		
Exterior Masonry Veneer/Siding/Trim - Specify:		$ 28,000		$ 28,000		
EXTERIOR WEATHERTIGHT COMPLETE	**Subtotal**	$ 66,000	$ -	$ 66,000	$ -	$ -
Interior Stairs		$ 2,000		$ 2,000		
Insulation*		$ 6,000		$ 6,000		
EIFS		$ 14,000		$ 14,000		
Drywall Finish				$ -		
Exterior Metal				$ -		
Cabinetry		$ 15,000		$ 15,000		
Finish Lumber		$ 3,000		$ 3,000		
Countertops: Marble/Tile				$ -		
Interior Doors		$ 2,000		$ 2,000		
Entry Door		$ 1,500		$ 1,500		
Finish Carpentry Labor		$ 4,500		$ 4,500		
Manufactured Home Delivered**				$ -		
DRYWALL / FINISH CARPENTRY COMPLETE	**Subtotal**	$ 48,000	$ -	$ 48,000	$ -	$ -
Finish: Ceramic Tile/Granite/Marble		$ 4,000		$ 4,000		
Interior Décor - Paint & Wall Covering		$ 9,000		$ 9,000		
Built-In Appliances		$ 8,000		$ 8,000		
Plumbing Finish (Including Fixtures*)		$ 16,000		$ 16,000		
Electrical Finish (Including Fixtures*)		$ 10,000		$ 10,000		
Finish Flooring - Hardwood, Tile, Carpet, Vinyl		$ 14,000		$ 14,000		
Heating/Ventilation/AC - Finish		$ 5,000		$ 5,000		
Bathroom: Tub/Shower Doors/Mirrors		$ 3,000		$ 3,000		
Finish Grading*		$ 3,000		$ 3,000		
Carpet		$ 2,000		$ 2,000		
Concrete/Asphalt Paving: Sidewalks, Driveway, Approach*		$ 4,000		$ 4,000		
Dumbwaiter		$ 5,000		$ 5,000		
Debris Removal		$ 8,000		$ 8,000		
Security System, Intercom		$ 4,000		$ 4,000		
Mirrors				$ -		
Final Cleanup		$ 1,000		$ 1,000		
Bath Towel Bars		$ 1,000		$ 1,000		
Contingency Reserve		$ 40,000		$ 40,000		
FINAL INSPECTION - BLDG. COMPLETE	**Subtotal**	$ 137,000	$ -	$ 137,000	$ -	$ -
TOTAL DIRECT COSTS		$ 419,000	$ -	$ 419,000	$ 30,300	$ -
FINAL DRAW - 10% RETENTION		$ 41,900	$ -	$ 41,900	$ 3,030	$ -
TOTAL PROJECT		$ 490,900	$ 30,000	$ 419,000	$ 30,300	$ 71,900

I hereby certify that the above referenced numbers are accurate and complete to the best of my knowledge.

X_____   _____
Borrower   Date

X_____   _____
General Contractor   Date

Figure 10-1:
An example of a draw request form along with the attached cost breakdown section on the left.

Courtesy of Stratford Financial Services

Managing the flow of funds with percentages and contingencies

Banks using the draw system are constantly managing the flow of funds on a *percentage of completion basis,* that is, the amount of your custom home that is finished. For the most part, however, they work with you if you want to move cash from one line item to another (say, for example, if you're under budget on one part of the project but need more money somewhere else), as long as your project is within a reasonable percentage of completion. Understand that their interpretation of *reasonable* may or may not match your interpretation. For example, if you stated that your framing was going to cost you $35,000 and it only cost you $20,000, the bank will probably happily let you use the extra $15,000 elsewhere in the build. However, problems may occur when the bank figures out that you've completed 50 percent of the house and you're requesting draws exceeding 60 percent of the budget. You may want to discuss the situation with the lender, and you may have to bring in your own money until the percentage is in line.

Sometimes you may disagree with an inspector about the percentage of completion on a particular item. For example, if you make a request for 50 percent of your framing dollars and the inspector claims it's 30 percent complete, you may want to protest because the lender may only give you 30 percent of the framing money. No one will go back out to the property to count the number of studs put in place. The bank will trust the general assessment of their trusted inspector. You can challenge the inspector's opinion but probably won't get too far in your argument. Accept the fact that you can't do much about it now, but know that when the project is 100 percent complete, you can get 100 percent of the cash devoted to that particular line item.

Having additional savings or the ability to borrow extra money can save time and headaches at times where the bank won't release all the money you need — and maybe even help you sleep a little bit better at night (and, as you discover while building your own custom home, you need all the sleep you can get!).

All construction loans have a *contingency* — a pool of money set aside in the budget for when costs go over what you expected. You can draw money from this cushion by request, but usually only in an amount in line with the current percentage of completion on the project. (We discuss the bank's approach to contingency in Chapter 8.)

Paying the subs

Paying subs fast and on a regular weekly basis has its advantages. This regular paycheck allows (and encourages) them to work harder, and sometimes it gets you a better price to boot. But remember: The contractor and subs don't have an agreement with the bank — they have an agreement with *you*. In their eyes, where the money comes from is *your* problem. Having access to cash through savings, credit lines, or family can keep you from sleepless nights or, worse, robbing banks when bank and contractor disputes seem hopeless. In our combined experience of working with thousands of custom home projects, the only projects that fall apart completely do so because they run out of cold, hard cash. For more reasons and methods of acquiring cash for your project, check out Chapter 7.

Dealing with deposits

Most construction lenders understand the need for advance deposits on materials and are willing to accommodate your needs. Many items, such as appliances, windows, doors, cabinets, and fixtures, require deposits months in advance of delivery, and deposits can sometimes add up to as much as 50 percent of the line item stated in your cost breakdown. Most of the time, your lender will insist on paying you for these deposits as a reimbursement for money that you have already spent — so you have to show them vendor receipts or cancelled checks as proof of payment. Depending on the lender, the overall percentage of funds disbursed when making draw requests may or may not include deposits. If your lender doesn't cover the deposits, it reimburses you upon installation of those items.

Using the draw system to pay for your log or kit home deposits

Log and other kit homes generally require large upfront deposits of up to 50 percent before the manufacturers begin to put the materials together for your home. Because the kit home manufacturer is supplying all the materials for the home, this expense can add up to a significant portion of your hard-cost budget. In addition, most manufacturers prefer to receive the rest of their money before, if not right at the time of delivery. Many construction lenders using the draw system have agreed to accommodate the needs of these manufacturers and wire deposits to these manufacturers in advance. In addition, the lender generally sends another wire to the manufacturer when the shipment is en route or when it receives confirmation of its delivery at the building site.

If you're building a log, manufactured, or kit home, we advise that you discuss this issue with potential lenders. Unless you have the available cash to advance to the manufacturer yourself, you may need to pick a lender that meets your manufacturer's needs.

Acing your finals — Final completion and final funds

Most banks retain 5 to 10 percent of hard costs in the LIP account until you complete the project. To get this money and complete your loan, you generally need to provide the following items:

- ✔ Certificate of occupancy and recorded notice of completion in applicable states (from the county or city inspector — see Chapter 15)

- ✔ Evidence of current homeowners' insurance (from your insurance agent)

- ✔ Final draw request (you and the contractor fill this out)

- ✔ Final progress inspection (from the bank inspector)

- ✔ Unconditional lien waiver and affidavit from your contractor (or from you, if you're serving as owner-builder) stating that you (or the contractor) have paid all material providers and subs in full

- ✔ Verification of no liens on property (from the title company)

If you have difficulty providing this information, then your construction loan can't be paid off or rolled to a permanent loan until you resolve the issues. You may face some penalties, and interest continues to accrue until you resolve the issues and pay off or roll the loan. We need to emphasize this point: You always need to try to resolve contractor and subcontractor disputes — especially minor ones — as soon as possible. If you have monetary disputes (most often major ones) that can't be resolved quickly, then put them aside and deal with them after you've finished the construction loan if you can. You can find information on contractor remedies in Chapter 11, and we discuss how to deal with projects in trouble in Chapter 20.

Figuring Out Who Gets the Money — You or the Contractor

Banks have different approaches for deciding who they pay their money to. Often, they leave the decision up to the folks running the project — you and your contractor. Ultimately, all the money that comes from the bank must either go to its proper place (vendors or subs) or directly to you or the contractor to reimburse you for cash that you've paid out yourself. Creating a system of payment — and including checks and balances — can help you keep your project healthy and happy (and that's the best kind of project to have!).

Suppliers generally want you to pay them immediately — or at least within 30 days — and they may require large deposits for certain items. And because subs often pay their laborers in cash, they may want you to pay them on a weekly basis. You and your contractor need to discuss each of the supplier's and sub's needs in detail to figure out how to handle the money flow. Make this discussion a part of your regularly scheduled meetings. You can find more information on managing money and having regularly scheduled meetings with your contractor in Chapter 11.

The bank may decide to control the funds. If the bank takes this approach, it most likely uses the voucher system discussed in the section "Understanding How the Voucher System Works," earlier in this chapter. More often than not, the bank simply asks you to designate a bank account to which you want the bank to wire the funds. In this case, you're the caretaker of the bank's money. You're paying interest on it and, if you mishandle it, you can lose your entire project (not a good thing!). You need to carefully consider how to control the funds properly. (For tips on keeping your project organized, see Chapter 2.)

Usually in a draw system, you get the choice of where the money goes. If the bank has left the decision of who gets the money up to you, you have three choices:

- **The money goes directly to the contractor's bank account.** If your contractor is carrying the entire financial burden, then she may request that the funds go directly into her bank account. In this situation, you need to make sure that the contractor has completed all the work and paid all the bills *before* you request the funds from the bank. This method gives you the least amount of financial control over your project, and we don't recommend it.

- **The money goes to a joint account for you and the contractor.** This method is very common. It assures the contractor that money is waiting in the account, ready to be spent on the project, but it doesn't place full control in any one person's hands. You need to set up the joint account with checks requiring two signatures (one from you and one from the contractor) before you can cash them. You and your contractor need to meet on a regular basis to discuss who gets paid, to make sure that the work has been completed, and then to issue the checks. If you travel extensively, or if the project is a second home in a remote area, you may need to make special arrangements to insure money flows in a timely manner.

- **The money goes into your bank account.** Obviously, this choice gives you ultimate control. In this case, you must set up a separate bank account specifically for the construction project. You're responsible to make sure that everyone has met his or her obligations on the project. If you make this choice, you need to be aware that the project's flow depends solely on your money-management skills. You need to have available cash at all times to make sure that you pay suppliers and subs regularly and keep them happy. You can then be comfortable knowing that the bank's funds come directly back to you.

Managing the books

Some people have difficulty dealing with large numbers and complicated accounting systems. A number of computer programs can help you out with that problem. Many people find Quicken useful. If figuring out a new program is challenging for you, you may find *Quicken For* *Dummies* by Stephen L. Nelson (Wiley) a useful tutorial in setting up your project. The bank that gives you the construction loan may have special offers for checking accounts that tie into online or software-based accounting programs.

Make sure that you have considered the possibility of a contractor change in the middle of the project. Although everyone hopes for smooth sailing, outside forces, including health, family, or quality of work, can cause a contractor to leave the job. You need to readdress whatever decisions you made regarding funds control. Discuss this possibility with your contractor and bank in advance so that you can make the necessary adjustments quickly and easily.

Using Credit Cards Responsibly Can Buy You a Trip to Europe

Regardless of whether you use the bank's money or your own, you need to manage cash flow accordingly when building a custom home. By using credit cards to pay building expenses, you can preserve your cash during your project. Yes, we know. Some people have nightmares thinking of racking up tens of thousands of dollars in credit card debt, but you only need the money for a short period of time.

Your credit cards can be valuable tools to use when making payments to subs and suppliers while you're waiting on your bank to release draws. Some of the benefits include

- ✔ **You don't have to pay any interest on most credit cards until the billing cycle has started.** You usually have up to 30 interest-free days to pay them back when your draws come from the bank.

- ✔ **Credit cards provide easy monthly statements for accounting.** These statements itemize exactly who you paid and when you paid them.

✔ **You can earn travel points.** Some people charge large amounts on their airline credit cards so that they can get the added bonus of accumulating frequent-flier miles. For example, American Express has point programs that you can use for travel or merchandise.

Unfortunately, not all your subs and suppliers accept credit cards, but you may be able to charge cash advances and still get some point benefits. Check with your existing card issuers to see what options may be available. Who knows, you may spend $200,000 in materials and rack up 200,000 travel points. These points give you enough for a much-deserved free vacation when you finish your house (and, after building your own custom home, you may just need that vacation!).

Although you may find credit cards useful after you've started the build, you need to wait until after the bank approves your loan before you open any new lines of credit or request credit increases to your existing cards. Every time someone looks at your credit, it generates an inquiry in your credit report. Too many inquiries in a short period of time can lower your *credit score,* which the bank uses in qualifying you for the loan. If you're financing your build, be sure to check with your loan officer before adding any additional credit.

Part III
Hammers and Nails: The Construction Process

The 5th Wave   By Rich Tennant

"We're gonna be late with the family room. The fellas started horsin' around on break and now they're into Act II of 'Peer Gynt', and you just don't rush Ibsen."

In this part . . .

It's time to build! Hundreds of people are now bringing their skills and labor to your soon-to-be doorstep to create your new home. In this part, we provide you with insight to managing the contractor and subcontractors without pulling out any hair. We walk you through the foundation, framing, and rough and finish parts of construction. We also give you spot-checks so you can check work along the way in case you notice something out of whack.

Chapter 11

All the King's Men: The Contractor and His Cohorts

*B*uilding a custom home requires many people working together in a coordinated manner to accomplish a series of jobs in an orderly fashion. Working with a construction crew demands communication and patience. In this chapter, we explain how to keep good communications with your contractor, subcontractors, and suppliers. We give you the skinny on contracts and discuss how to resolve disputes. We also give you the inside scoop on working with inspectors. Finally, we point out the risks and remedies with mechanic's liens.

Working with Your Contractor

A large part of success in working with your contractor is picking the right one. Just because a contractor offers a good price and builds good homes doesn't mean she is right for your project. In Chapter 2 we provide plenty of good information on selecting the right contractor for your project and also understanding the contractor's role.

To keep everything working effectively with your contractor, you need to be proactive. Make sure you're clear and decisive about what you want and more importantly what you don't want. Ambiguity can cause mistakes and delays. If you aren't sure of what you're saying, then ask more questions or do some research. Your contractor will also appreciate clarity in your conversations. The last thing she wants is to fix a misunderstanding.

Impatience is your enemy. A custom home project is organic in nature and doesn't move in a precise fashion. For example, you may think the framing should be finished in three weeks, but a number of delays, such as the weather, scheduling errors, or manpower shortages, can make it take four weeks. Relax. Take a deep breath. The project will develop its own rhythm and eventually it will be completed. Although you do need to be working to keep your house on schedule, you don't want to be constantly rushing everything. Contractors and subs are human and will work poorly in a high-pressure environment. Settle in for the long haul and have faith in the professionals you hired.

Fostering good communication — A meeting a day keeps the anger away

Hands down, the surest way to keep a custom home project progressing is with steady, effective communication. Err on the side of more communication rather than less. Everyday something new is happening on your project, so the easiest way to keep everyone informed is to have a meeting before the work starts.

We realize that everyone thinks meetings are boring and having one every day sounds like a huge waste of time. It doesn't have to be. Think of a meeting as two-way communication that can happen by phone, e-mail, or in person and that only lasts a few minutes. This daily update needs to address the following five basic issues:

- What is happening on the project today?
- Are we on schedule?
- Are we on budget?
- What problems or issues, if any, are occurring on the project?
- What, if anything, do you (the contractor) need from me (the owner)?

If everything is running smoothly, you and your contractor can proceed assured. If you and your contractor notice certain issues, you can be confident in having a time everyday to discuss and resolve them. Having this regular meeting time saves panic calls and work stoppage while you're busy earning money. If you and the contractor both e-mail, you can swap the answers to the questions the night before and agree to talk only if there are any problems.

For good communication, both sides have to be proactive. Be clear and honest on all your issues. Remember that fear goes both ways. If you're dissatisfied or confused, bring it up. How does your contractor know that something on your end is a problem unless you speak up about it? Believe us, she doesn't!

Maintaining a productive work environment

Before you're ready to start work on your new home, you and your contractor need to have a preconstruction meeting to set reasonable expectations on both sides for a productive work environment. This meeting is your chance — and the contractor's — to remove miscommunication before it happens. It also lets everybody know what the hot spots are that can cause problems later. Use this list to figure out what you need to discuss and agree upon:

- ✔ What are acceptable communication times?
- ✔ What is reasonable contact during work hours?
- ✔ How often will we address the accounting?
- ✔ Will the property be fenced off and locked?
- ✔ What is a reasonable level of cleanliness on the site?
- ✔ Will smoking be allowed on the site?
- ✔ What potential noise restrictions may be an issue?
- ✔ What will be the working hours on the site?
- ✔ Will workers be restricted from removing shirts?
- ✔ What bathroom accommodations will be set up for the workers?
- ✔ Will workers be aware of using foul language?

These issues commonly impact the working environment of a construction job site. Make sure you address them with your contractor.

Before you and your contractor set agreements on these issues, talk to the neighbors and solicit their input and approval because they'll be impacted by some of the decisions even more than you. Also check to see if any local government restrictions apply. Taking these proactive steps can help you develop better relationships with your neighbors and make for a much more pleasant move-in process.

Change orders — Dealing with indecision

If one phrase causes a contractor to wake up screaming in the middle of the night, it's the dreaded *change order*. A change order is needed when you decide to do something different than you agreed upon in the plans or contract. You're entitled to make any changes you want. After all, you're going to live in this home. However, change orders can create problems and delays not to mention added costs.

Try to keep change orders to a minimum if at all possible. Doing so can be difficult because your house will look different as you build it. For example, a room may seem too small or a window may need to be farther over to catch the sun. Minor change orders, such as moving a small wall or adding a door, may not cause too much trouble, but major change orders, such as adding a room or changing a structural wall, could send you back to the planning department and cause weeks of delays.

Plan ahead and use a change order as a last resort. If you do need to make a change, tackle it early and immediately. Indecision and constant change can disturb any project's rhythm. A long delay can upset every sub's schedule. Furthermore, the contractor usually isn't willing to absorb the costs involved in change orders and may even charge a premium as spelled out in the contract. Don't let the pace of the project lose any momentum if possible in order to keep extra costs to a minimum.

Keeping a happy and healthy relationship with your contractor

The partnership between you and your contractor will be the key to the success or failure of your custom home project. You'll probably share more issues, problems, and emotions in one year than most couples deal with in a lifetime. Utilize these five keys to maintain an excellent working relationship with your contractor:

- ✔ Do your homework first, and then trust the contractor to do his job.

- ✔ Detail every decision and ask every question before you start. Imagine you're going out of town for the entire project and the contractor could only work from written instructions.

- ✔ Let the contractor know your fears before you start so he can address them.

- ✔ Ask that the contractor not start the project until he can commit to working continuously through to the end.

- ✔ Think of your contractor as being on your side, and he most likely will act like it.

Managing Your Time and Money

Any business-related project boils down to managing time and money. A custom home project is no exception. Your project isn't a charity project. Your contractor, subs, and suppliers build homes for a living. They want to make their money and do it in as little time as is reasonably possible. They want everything handled in an efficient professional manner — they work best when they don't have to chase their money, and everything happens according to expectations and agreements.

Of course, living up to commitments goes both ways. Check in regularly with your contractor to make sure the project is running consistently with the budget. If the project is running over budget, go over the details of the expenses so far to determine that the contractor is living up to his obligations. If you have a problem, you can refer to the "Managing Disputes" section, later in this chapter.

Executing contracts

Whether you hire a contractor or act as your own, you'll have to execute contracts with the people working on your custom home. If you hire a contractor, he will have contracts with the subs, so you'll only have to negotiate with him. (See Chapter 2 for more about your contractor's specific roles.)

Contracts between you and the contractor spell out everything in writing with a bunch of legal gobbledygook in case a dispute occurs. If all goes well, you'll put the contract between you and the contractor in your file after signing and never look at it again. If not, it's a useful tool to remind both you and your contractor exactly what you agreed upon. A good contract is the basis for good communication. You can start with a copy of the contractor's standard contract or a contract he used with a previous client if he doesn't use a standard contract.

Contracts are legally binding documents; you need to read and understand them thoroughly before signing anything. Make sure you understand the project's price, payment plans, and your obligations. Have everything in place on your end to make good on your side of the deal. Doing so can be difficult if, for example, your construction loan isn't approved. One solution for this issue is to make the contract contingent upon financing.

Do your best to negotiate win-win situations on each point and make sure you have your attorney review the contract before signing on the dotted line. The contract is always subject to change provided you and your contractor both agree to make the changes and initial them. A good contract includes the following list of required items:

- ✔ Names, addresses, and phone numbers of all parties named in the contract
- ✔ The contractor's license and type
- ✔ Tax ID or Social Security number
- ✔ Workers' compensation insurance number

- ✔ Job address
- ✔ Job specs
- ✔ Price and terms
- ✔ Time frame
- ✔ Responsibility for delays
- ✔ Terms of warranties
- ✔ Terms for conflict resolution or arbitration
- ✔ Signatures and dates

One question Kevin is asked all the time is whether to choose a fixed-price contract or a time-and-materials contract.

Fixed-price contracts

A *fixed-price contract* means you agree on a single price for the project, and the contractor makes his profit on the difference between that price and the total of labor and materials. Many lenders require fixed-price contracts because these contracts protect them from cost overruns. Some contractors believe, however, that fixed-price contracts create an adversarial relationship between the consumer and the contractor. The contractor has to look to cut costs in order to make more money, which may take its toll in quality on your house if costs are running higher than expected. You can add language that allows for renegotiation should cost increases exceed a certain percentage.

Time-and-materials contracts

With a *time-and-materials contract,* the contractor bills you at her cost for the labor and supplies and then makes a percentage profit based upon the costs. This type of contract solves the problem of the contractor looking to skimp in order to make more money; however, some conflict may still arise because the contractor can make more money by making you spend more.

Kevin recommends creating a variation that offers additional incentives for bringing the house under budget and on time. Doing so can give you the best of both worlds, but your lender needs to agree to this type of contract.

Check your local laws regarding deposits. Often contractors ask for a security deposit with the contract. Asking for a deposit is legal but some states don't allow large deposits to contractors. If the contractor is asking for more than 10 percent, check the limits with the state contracting licensing board. Be aware that any money you put on deposit is at risk of being lost until the project actually starts.

Scheduling the project

Your contractor creates a schedule for the project based on the following four elements:

- The plans' requirements
- Subcontractors' availability
- Supply availability
- Weather

Your contractor does his best to keep to the planned schedule, but unexpected changes in any of these four areas can throw everything off kilter. At that point he adjusts and makes up time wherever possible without sacrificing the quality of the house. Both of you want to keep the project moving everyday. Both you and your workers will stay happier if you see some progress at the end of each day. Stagnation can break emotional momentum and create a dismal working environment.

Controlling the funds — How to manage the checkbook

Whether you finance the property as we discuss in Chapter 8 or decide to build using all your own cash, you'll need to establish a method of getting everyone paid. If you're paying for the project out of your own pocket, you can just write the checks yourself every day or fund the contractor in stages. If you're getting a loan, you have to designate who gets the money, or the lender may designate for you. You have three basic options, which we touch upon here and cover in greater depth in Chapter 10:

- **You control the funds.** You have total control of the money, but you must sign every check and make sure people get paid when they need their money, which can be inconvenient.
- **The contractor controls the funds.** This method is much more convenient, but the contractor now has full responsibility for your money, which could be risky.
- **You both control the funds.** A joint checking account gives both of you access and control. You can set up the account where both you and the contractor have to sign checks on big-money items. Using a joint account enables you to watch the larger expenses while giving the contractor funds for day-to-day expenses.

Either way, your best bet for running the project smoothly is to have funds readily available as we discuss in Chapter 7. The lender-disbursement systems explained in Chapter 10 may not be consistent with the needs of the contractor and subs, so extra funds act as a cushion from bank bureaucracy. You want to make sure everyone on the project is happy and not begging to be paid. Some subs may ask for advances for materials or to secure their time. Try not to pay any more in advance than you have to, but don't leave them hanging either.

Many subs and suppliers offer you better deals if you actually pay cash rather than writing a check or charging your credit card. Cash is better for them if they want to avoid taxes by not reporting the income. Because income reporting is their responsibility, you need to decide on your own conscience and moral view of this issue. Using cash to pay isn't illegal, and it can save you money (from credit card interest), but handling large amounts of cash can be difficult and can make you feel a little dirty if you suspect that the people you are paying are cheating on their taxes. Remember that canceled checks and credit card statements serve a useful purpose besides their ease of use: They are your receipt that someone actually got paid. The fact that cash is undocumented can work against you as well.

Introducing Other Important Players

Aside from your contractor, you work with many other people on the project. Your contractor has the primary communication with subs, suppliers, and inspectors, but you'll be involved in many of the conversations. If you decide to be an owner-builder, you'll have to manage all the communication. (We explain the advantages and disadvantages of being an owner-builder in Chapter 1.)

Dealing with suppliers

Suppliers can range from a Home Depot store, a specialty materials store that caters to builders, and even individual artisans that make fixtures for homes. Each one has its own terms for deposits, payments, and delivery. If you're working with a contractor, he'll probably arrange for most of the basic materials from a supplier with whom he has an established business relationship. (See Chapter 2 for the specifics on a contractor's role.) Owner-builders need to find their own suppliers — local lumberyards and the Internet are a great place to start.

Even though your contractor provides the suppliers, you can still look to buy fixtures and items. You may find unique items or save money if you have some connections. The Internet is a great place to shop, and stores like Lowe's or Home Depot Expo can be great resources to buy exactly what you like. If you buy items, then don't make anything more difficult for your contractor. To make your contractor's job a little less hectic:

✔ Communicate as soon as possible to make sure she hasn't already ordered the item.

✔ Make sure the item is practical for the project. For example, older electrical and plumbing fixtures may require some adaptation for modern usage.

✔ Make the item easily available to the contractor by having it delivered to the site.

✔ Ensure the finances have been reconciled with the budget and contract.

Suppliers often need deposits to hold materials and for large items ordered in advance. Make sure you have plenty of cash available for deposits and to keep the suppliers happy. See Chapter 7 for more about managing your cash flow on the project.

Working with subs — Each one is an expert

Subcontractors are contractors too. They may not have all the credentials of a general contractor, but they're experts in their specialized fields. They're referred to as subcontractors or *subs* on your site because they work for the general contractor. Which subs your contractor hires depends on your project.

The following is a list of subs probably necessary for your project in the approximate order they will work on your project:

✔ Grading sub or excavator

✔ Foundation sub

✔ Concrete sub

✔ Framing sub

✔ Plumbing sub

✔ Electrical sub

✔ Mechanical/HVAC sub

✔ Roofing sub

✔ Drywall sub

✔ Painting sub

Your project may require other specialty subs. Subs may also handle items, such as tile, masonry, millwork, stucco, insulation, sewer/septic, and carpentry. Most of these subs have licensing in their specific area. Ask your contractor to verify their licensing before they start work on the project.

Your contractor performs most of the communication with the subs, especially in the early structural phases of the build discussed in Chapters 12 and 13. Make yourself available to walk through the property with the primary subs with whom you may have some input, such as:

- The plumbing sub
- The electrical sub
- The HVAC sub
- The roofing sub

By walking through with the subs before and after their work, you can make sure you clearly communicate about everything important to you, such as the locations of vents, fixtures, and switches. The finish work (see Chapter 14) requires many more individual walk-throughs and meetings to be assured that everything is kosher.

Subs are craftspeople first and businesspeople second. Many of them are less concerned about their social skills than their work quality. You want quality to be their priority — but the craftsmanlike manner of subs can also make for difficult communication. You need good communication and rapport during the process, especially for the finish work. Do your part to create a warm friendly environment. Say hello to the subs when you're on-site. Stop by with a cooler of soft drinks on a hot day. A little good can go a long way in building a good rapport with the workers. Use your contractor to manage most of the work-specific communication — let him be the bad guy when necessary. That's what you're paying him for.

Preparing for building and bank inspectors

Two kinds of inspectors check in to see how your project is going:

- The **bank inspector** determines how far along the project is so you can be reimbursed for the work that's been completed. These inspections occur at specified times based upon the draw schedule discussed in Chapter 10.
- The **building inspector** visits to make sure that the project is being built to meet the local codes and according to the approved plans. These inspections occur based upon local government requirements established by the local building department.

These two inspectors aren't interchangeable and have no interest in looking at anything beyond their scope. Contrary to popular belief, the inspectors are absolutely on your side. They both want the house to be completed in a reasonable time frame and in a workmanlike manner.

In order to help the inspection process go as smoothly as possible, you and your contractor need to use these few tips:

- ✔ Call the inspector a few days in advance so you don't have to stop the project and wait.

- ✔ Keep the site clean and organized so the inspector feels good about what he sees.

- ✔ Make it easy for the inspector to see what he needs to see so he doesn't have to spend time digging and become frustrated. For example, don't stack the appliances in such a way that they hide the inspection areas.

- ✔ Keep disputes calm. Find out how to fix the problems and only pick battles you're sure you can win.

A good relationship with inspectors can keep the project moving and running smoothly. Keep everything cordial by smiling and saying "thank you" a lot. If you have a problem, take a step back and communicate with your contractor about how to get everything back on track. Ultimately, you may be right in an inspection dispute, but inspectors have the power to shut everything down. Find a way to see their point of view and go with it.

Managing Disputes

Because you have so much at stake in your custom home project, emotions can run high. You can quickly resolve most disputes after tempers have settled down. Usually everyone on the project has the same objective of getting the house complete —your satisfaction with the final product. Most problems relate to your dissatisfaction with the work quality or a cost issue. Remember that in a dispute situation you're ultimately responsible for the outcome. Even though the bank has a financial interest, it will take the path of least resistance — which may not be in your interest. Never will the lender act as your advocate. When something goes sour, you can be assured that the parties will be looking out for their own best interests.

Avoiding mechanic's liens —
The contractor's weapons

A *mechanic's lien* is a debt instrument that attaches to your property and guarantees the contractor will get paid before anyone else does when the house is sold. A mechanic's lien takes precedence over all other liens, including your mortgage, which makes it a very powerful tool. A contractor can file a mechanic's lien anytime up to 90 days after the house is finished regardless of when the work was actually done.

Because of the power of mechanic's liens, lenders and title companies charge you extra money and make you sign indemnifications to protect their interest and put the weight of responsibility back on you. If a mechanic's lien is filed against your property, you'll have difficulty getting money from your construction loan, and you can't roll the mortgage into permanent financing.

 Because you bear the burden when a mechanic's lien is filed, you need to avoid them at all costs. You're better off paying a contractor and suing him after the fact rather than withholding payment over disputed work. You may think you have leverage, but the penalties for a loan that never rolls to permanent financing, as discussed in Chapter 10, can far exceed the original contract amount in dispute.

To protect yourself against mechanic's liens, use the following tips:

- **You can require the contractor to supply a performance bond.** This bond completes the project and/or pays damages up to the bond amount. This money can pay off any outstanding payments due. These bonds can cost 1 to 5 percent of the construction contract and need to be recorded with the county.

- **You can use a joint fund control.** Local services that manage the books and take responsibility for paying all subs and suppliers are available for a fee. These services can provide lien waivers and other protections.

- **You can pay with joint checks.** Joint checks have both your name and the contractor's. Using joint checks helps to document to suppliers and subs that you have met your obligations. Regardless of whether or not you adopt any of these methods of protections, you absolutely need to collect *Waiver and Release* forms. Have every sub and supplier sign these forms at the time you pay them, and make sure your contractor collects them as well. You need these forms for your lender at the end as well. You can get these forms at a stationery store.

Using legal remedies — Arbitration and attorneys

Your contract needs to spell out the details of how to resolve major conflicts. If you and your contractor can't reach an agreement, then you may have to resort to legal means or at the very least binding arbitration.

Pre-liens, nothing to be afraid of

Part of the process of filing mechanic's liens requires that subs and suppliers must send you a notice that work was done or materials were provided to the project. These pre-lien notices insure that you're aware of moneys owed to people that you haven't seen working on the project, such as subs hired by the contractor for one day's work. Don't panic when you receive this notice. A mechanic's lien itself only occurs when it's filed with the county recorder. Keep these documents and obtain lien releases from each of the suppliers and subs when the project is finished and they have been paid in full.

Arbitration is an alternative to the courts that can save time and money. Instead of hiring attorneys, the two parties present their case to an independent arbitrator who examines the situation and works to find an agreeable solution. If no solution is agreed upon, then the arbitrator creates one based on the information reviewed. This process can take less than one-third the time and cost one-fifth the amount of a court case. Many contracts specify the American Arbitration Association (www.adr.org). Both parties agree that they'll use arbitration in lieu of legal action.

If arbitration isn't specified, then the last resort may be to engage an attorney. You must assume that if your project has gotten to this stage, then you are prepared to spend tens of thousands of dollars and possibly years of time before a resolution happens.

Consider these four bits of information before you engage in a lawsuit:

✔ Do your best to confirm that the party you are suing has the means to pay a judgment or at least has insurance.

✔ Be sure you truly have legal ground to stand on and you aren't acting on an emotional basis.

✔ Be sure you can absorb the costs associated with a trial and any delays to your project even if you lose.

✔ Be sure you have the emotional strength to deal with attorneys and personal attacks.

If you're unsure in any of these instances, you may want to head back to the bargaining table.

Not all attorneys are well versed in construction law. Kevin has seen fee-hungry general practitioners cost their clients tons of money in cases that never should have been filed. Many attorneys believe they're right even when they're wrong. Make sure your attorney is an expert in the field and can keep the emotions out of it. Stick to the facts and be ready to admit you're wrong if you are.

Chapter 12

Excavation and Foundation: Getting a Solid Start

*T*he design is finished, and you've paid for all the permits. (If you haven't paid for your permits or have any questions about the permit process, check out Chapter 6.) You've also picked a contractor, and you're ready to get to work. This chapter provides you with a basic understanding of the first phase of the actual construction process — excavation and foundation.

As you progress through this phase, you and your contractor need to meet regularly to keep track of the progress and the schedule for all the subcontractors to complete their work in perfect synchronicity (see Chapter 11 for more on working with your contractor and subs).

The first steps in the construction phase aren't for do-it-yourselfers because they require heavy equipment and complex engineering that few people are skilled in. More than a few custom homes have gotten into trouble when their happy owners underestimated the complexity of these first steps, particularly building a foundation. Our goal in this chapter is to be sure you understand just how important this part of the project is — your home literally depends on a firm, well-built foundation for its survival — and your role in ensuring that you get the best foundation possible.

For us to give you every detail of construction in this chapter would be impossible, so instead we focus on the important stuff — the vital information you need in order to have meaningful discussions with your contractor and subs. Most of you depend on your contractor for the tasks in this chapter, so we include spot-checks you can use to make sure everything runs smoothly. You probably won't have time to check every detail, but these lists can alert you to a problem or give you the confidence to move forward.

Surveying and Site Preparation

In order to start, you need to know where everything goes and have a clean site to work with. In this section we discuss how everything finds its rightful place and how to make the ground ready for your home's foundation.

Using your survey

Before you dig in, so to speak, make sure you're digging in the right place (don't laugh, digging in the wrong place happens more often than you think). It's time for a survey. No, not the kind they do on TV's *Family Feud;* this survey requires a specially trained professional (a *surveyor*) to use surveying equipment to mark the following:

- ✔ **Lot lines:** These are the property boundaries.
- ✔ **Setbacks:** These tell you where your building limits are relative to your neighbor's property.
- ✔ **Underground utility lines:** If your property has any power lines and water pipes buried on it, identifying their location can keep you from hitting them.

When your surveyor completes the job, she places a numbered tag on a permanent stake on the property. This tag has the surveyor's registered license number and serves as a reference point for all future surveys. Often the survey is completed early in the design process, particularly if you're dealing with a large rural lot where the boundaries aren't clearly marked or if they're required by the building department. You, your contractor, or your architect can hire the surveyor, depending on when in the process she is needed. The cost varies widely by region and the size of the lot.

Designating the lot lines

First, your surveyor comes out to your building site and marks the lot's borders. In fact, the surveyor's most important job is to carefully mark out each border.

Your surveyor identifies the lot lines by using a tool called a *transit,* which looks like a telescope (no, you can't use it to see Saturn's rings at night). She finds permanent tags or markings previously installed by the government that correspond to the *plat* map (the official, recorded map that shows property boundaries and measurements) provided by the county. You'll receive a copy of the plat map from the title company; you can also easily obtain it from the county because it's public record.

Then, using the site map, the surveyor and contractor preliminarily mark out all the important areas (the corners of the house and garage, the pathways for utilities to be brought into your home, the edges of the driveway and walkways, and so on) with stakes, spray paint, and ribbons. Marking them now makes it easier to put in the more permanent markings discussed later in this chapter in the section "Marking the build site."

Looking at setbacks

Most cities and counties call for some separation between houses, called a *setback,* of some distance in feet, which your architect will have included in your plans in order to get them approved. We discuss setbacks in more detail in Chapter 5.

If you're building on a large lot, you probably won't be encroaching on these setback limits, unless your home is built close to the property line. On a small lot, however, you have a much greater chance that you could encroach on a setback, leading you directly to a major headache. Why? Because, if indeed you have encroached on a setback, your friendly, local building inspector could force you to remove the work you have already done and start again. And if you're well on your way in the construction process, having to start over again could cost you significant amounts of both time and money!

Keeping your ear to the ground for utility lines

Have your surveyor and contractor not only mark the underground utility lines but also flag them in case the paint markings are lost due to excavation and erosion. Make sure the utilities remain marked during the entire earth-moving project. For obvious reasons, your neighbors-to-be probably wouldn't appreciate it if their power, phone, gas, or water service were accidentally cut off for a day or two when one of your subs ran a backhoe through a buried utility line.

Before any earthwork begins, have your contractor call your regional organization designed to protect accidental damage to underground utilities. Each area has a phone number to call as a free service. (You can call 800-258-0808, or you can access your local number by logging on to www.electricsmarts.com/content/callbeforeyoudig.asp. The Web site also has a color chart for marking locations.) Most utilities cooperate with this service in order to protect accidents form happening. Often the utility company comes out and marks in colored chalk where the utility lines exist. Taking this step saves the utility company money and can save lives as well.

If you *don't* call and you hit a line during the course of construction, the penalty fines can be as much as $100,000 plus damages. Do you have that much spare change just lying around? (For more on hooking up the utilities, check out "Preparing for the Utilities" section, later in this chapter.) Finally, you or your contractor need to call the utility companies to relocate their services if they need to be moved to accommodate driveways, decks, and landscaping, because these items could block future access.

Preparing your site before the first shovelful

You're probably anxious to break ground, but not so fast. You want to make sure everything is ready, and you need to take several steps before anyone digs out the first shovelful of dirt. You and your contractor need to discuss the benefits and costs for the following suggestions before you implement them:

- ✔ **Post prominent "Danger" signs (in red) throughout the building site.** Doing so can help protect your liability in case someone unauthorized might trespass on the property.

- ✔ **Arrange for a rental toilet for your workers.** Everyone has to go sometime and, more importantly, somewhere. Believe us: The alternative isn't very appealing.

- ✔ **Decide whether or not to install an onsite storage shack for materials.** Not only can you keep your materials and valuables dry in an onsite storage unit, but you can also lock up any valuables for security purposes. Installing a storage shack is a good idea in a moist climate, but probably not necessary in places where rain or snow is infrequent. The storage shack doesn't need to be fancy. In fact, an aluminum shack from a hardware store may work just fine. If your project is a big one, however, consider renting an industrial shipping container. For just a few bucks a day, you can have a watertight storage facility delivered to your site at the beginning of your project and then picked up at the end.

- ✔ **Consider adding an administrative office for your contractor.** You may consider putting a trailer or motor home on the lot, particularly if you plan to spend a lot of overnights on the property as construction progresses. Adding an office can also be a very convenient place to store paperwork, tools, and refreshments as well as to provide comfort and first aid for workers when necessary. You can then post the building permit on the trailer or motor home's door (if you don't have an office, then you can simply post the permit to a post on your property or on your new home's framing or sheathing when it is far enough along).

- ✔ **Discuss with your contractor about putting up a lockable chain-link fence.** The fence can keep out kids and other intruders after hours. Installing a fence not only protects your site from theft and vandalism, but also prevents someone from coming onto the site and getting hurt. Be sure to look into renting the fencing instead of buying it.

Dealing with trees

Your site plan will detail which trees to save and which trees to remove. You make these decisions during the planning process, which we discuss in Chapter 5. Removing large trees requires a subcontractor with a bulldozer or tractor. The sub first cuts down the trees, leaving roughly 4 feet of trunk exposed, which is enough trunk to connect a chain for the tractor to pull out the entire stump. Don't leave stumps in the ground because they'll rot and attract unwanted pests as well as leave unsightly (and possibly dangerous) sinkholes in their wake.

The wood from the trees and stumps cut on your property is yours to keep, so talk to your contractor about putting it in a corner of the lot where you can cut it into firewood later or ask your contractor if he can have a laborer cut it up (the tree cutters will charge you dearly if you ask them to split the wood). If you don't have the room to store all that wood, consider selling or giving some to your neighbors.

Do you have some trees that you want to protect? If so, you have several different options:

- Have your contractor place hay bales around the trees to protect them from tractors and heavy equipment (and to be sure no one accidentally cuts one or more down). Later, you can use the hay to cover mud and to dry out puddles.

- Surround the trunks with bright orange plastic fencing from your hardware store. Doing so can protect the tree somewhat from flying objects and make it visible so heavy equipment and truck drivers will avoid hitting them. Make sure the fencing is tall so it protects the bark from being knocked off the tree, helping to protect the tree from the elements and pests.

- Tie red tape or ribbon to the trees. The bright color alerts workers from hitting the trees or accidentally cutting them down.

Clearing and grading

Depending on where you decide to build, your lot may also need to be cleared of brush and shrubs. Your contractor can arrange for the clearing. This removal can usually be done with a bulldozer at a reasonably low cost. After the lot has been cleared, grading can take place. *Grading* is scraping the land to move dirt from one place on the lot to another. Your contractor brings in a grading subcontractor to do the job. Tractors, bulldozers, and other heavy equipment often grade the land, although small grading projects are done by hand. The grading is completed in two stages:

- **Rough grading** redistributes dirt on a lot to create level ground at the building site (which is important, if you want to have a level house!). Any leftover topsoil may be pushed to a corner of the lot for later use in landscaping.

- **Finish grading** contours the yard in a way that is both aesthetically pleasing, and that moves water away from the house instead of toward it. (We address finish grading more in-depth in Chapter 17.)

When grading, make sure your contractor or sub extends the grading work at least 3 feet beyond the expected perimeter of the house and garage.

If your lot is rocky or has boulders, your contractor may have to arrange for *blasting*, a process that (at least in our minds) conjures up early morning cartoon images of a crazy coyote and some dynamite. Actually, those images aren't far off. Blasting can be dangerous and expensive, but it may be necessary to make the lot usable.

Constructing retaining walls

If your lot is on a slope or hill, it may require some serious grading to create a flat *building pad*. A building pad is the flat ground where your foundation will sit. Your engineer has calculated the best structure for the particular needs of your slope. For example, one alternative includes building several retaining walls in succession to create multiple flat pads for the house and yard, which is called *terracing*. The ground will be terraced or cut flat into the hillside.

Furthermore, in severe situations, the contractor may build a retaining wall to prevent the land (and your house along with it) from slipping. A general rule: Your contractor or a sub can build any retaining wall 3-feet high *dry*, that is, without using cement or mortar. Higher walls need to be built *wet*, using cement or mortar. Your plans will specify the necessary structure for the retaining wall. Another important rule to make sure your contractor knows: A retaining wall must extend 2 feet into the ground for every 1 foot of height above the ground.

Depending on where you live and your contractor's preferences, your retaining walls may be made from different products, including the following:

- **Cement blocks filled with reinforced steel and concrete:** These are common in the West where earthquakes are common.
- **Large logs or railroad ties:** The logs or ties have been chemically treated (often with *creosote*, or other rot and insect repelling materials) and bolted into place.
- **Keystone or boulders:** These are commonly used in the Midwest.

Your contractor and/or engineer can place your retaining walls

- Above the building pad to keep the slope from sliding down onto the house
- Below the building pad to keep the house from sliding down the hill

Check out Figure 12-1 to see both examples.

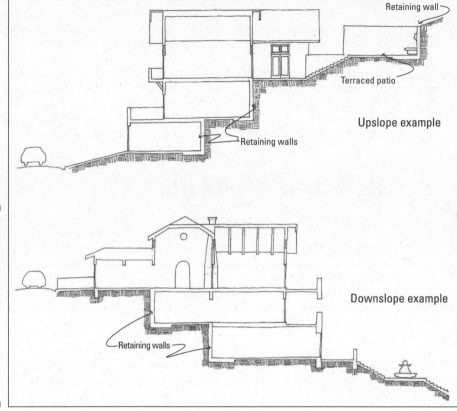

Retaining wall

Terraced patio

Upslope example

Retaining walls

Downslope example

Retaining walls

Courtesy of Tecta Associates Architects, San Francisco

Figure 12-1:
An upslope
retaining
wall creates
a terraced
pad below.
A down-
slope
retaining
wall creates
a terraced
pad above.

Providing drainage

Drainage is a very important part of the build process. In fact, approximately 80 percent of foundation problems (leaks, cracks, uneven settling) in houses can be later traced back to poor drainage. So, to avoid these kinds of problems (and to ensure your front yard doesn't turn into one big puddle every time it rains), your contractor needs to follow the engineer's recommended plan for proper drainage on the property. This plan is in the working drawings, which we discuss in Chapter 6.

To create good drainage, parts of your property will be graded with a slight slope (away from your home), encouraging water to run off the property.

If your house is going to have a basement, the contractor may need to install a *French drain* system. A French drain is a series of trenches dug around the house's perimeter, designed to direct water away from the foundation. To work properly, the trench must be located deeper than the desired depth of the basement. After the trench is dug, a polyvinyl chloride (PVC) pipe with holes in the top — run at a slight, downhill angle — is placed at the bottom of the trench and covered with gravel and then dirt. If desired, water from the drains can be directed to a collecting area where an electric sump pump moves the excess water to a surface drainage channel.

On some properties, PVC pipe, galvanized pipe, or corrugated plastic tile may be installed to direct water from specific areas on the property and toward other areas of the property away from the house. Most homes today are designed with a *closed system,* where all the water runoff from the house is channeled directly into these pipes and carried off the property.

Marking the build site

The next step is to clearly mark the building site. Using tape measures and the survey, your contractor marks the exact points for the edge of the house and garage. Depending on the type of foundation to be installed, your contractor may mark the perimeter with wooden stakes and boards called *batter boards.* The batter boards may only be a few feet long, but they'll clearly mark the corners as well as the height of the foundation to be installed. The contractor then attaches string between the batter boards to represent the sides of the foundation and ensure straight lines and square angles.

When the foundation area is clearly marked, then the contractor can start lining up utility hookups and getting ready for excavation. Your contractor will also need to stake out all the utility hookups before construction begins.

Spot-check — Surveying and site preparation

Everything is marked and cleared. Now is a good time to perform a spot-check to make sure you're ready for the next steps. No special tools are necessary here.

❏ Lot corners and setbacks are clearly marked with flags or paint.

❏ Separate areas for materials and trash are set away from the build site.

❏ "Danger" signs are clearly marked.

❏ Trees to be saved are protected and easily seen.

❏ Retaining walls are secure and ground is packed.

❏ Building site is clearly marked and batter boards are secure.

Discuss any issues or discrepancies with your contractor and determine when and how to fix them.

Preparing for the Utilities

You don't really appreciate utilities — electricity, gas, telephone, cable television, and sewer and water — until they're gone. And, although some people are generating their own energy (through solar, hydro, and wind power), pumping their own water out of the ground, or having their utilities delivered by truck (propane and fuel oil), most people rely on utilities that are provided by their local communities.

Some utilities come into your house underground (such as water and sewer) and some overhead (such as electricity and cable television, except in communities where these utilities are also underground for aesthetic reasons). Whatever the case, your contractor marks where all these different connections are coming from and going to according to the plans. You and your contractor connect with all the appropriate parties so that you and your construction crew can have the benefit of modern utilities during the building process — and beyond.

Connecting water

Depending on where your lot is located, you'll either connect to public water or have to put in a well if no public water is readily available. Public water is generally more cost effective and reliable, but rarely available in rural areas.

If you're connecting to public water, your contractor will have laborers dig a trench to the water supply, connecting to the main water pipe at the street. Your contractor installs a temporary meter so the water company can measure your water usage during the construction process. If the main pipe has a pressure of 80 pounds per square inch (PSI) or higher, you'll later need to install a regulator to reduce the pressure, avoiding excess wear and tear on the pipes (or unscheduled bursts!) in your house.

If you're installing a well, you or your contractor will arrange for a well company to install the well. The well company drills down into the water table until it can find clean ground water. After the well company digs the hole, it installs an electric pump to bring the water to the surface for home use. You'll also have a water storage tank so that your pump doesn't have to run every time someone flushes the toilet or takes a drink.

Regardless of whether you're using well water or public water, because water is generally used during construction, your contractor needs to install a temporary spigot for almost immediate access.

Hooking up to the sewer

If you're connecting to a public sewer system, your contractor will have laborers dig trenches to lay the drainpipes — generally made from cast iron or PVC — leading away from the house. These pipes need to be installed prior to any concrete being laid, and your site plan will determine the most effective route (see Chapter 6 for more info on site plans). Your sewer pipes need to slope downwards so that water drains toward the public sewer system.

Be sure that your sewer system has an antireturn valve to protect you from sewage flowing back into the house if your local sewer system overflows in a rain storm. Kevin's 1912 house had this problem and, besides creating a smelly mess that only the Creature from the Dark Lagoon could love, it was very costly to clean up.

Installing septic systems

If your property calls for a septic system based upon the work you did in Chapter 5, then you need to install one now. Your contractor consults the plans to determine the exact location of the septic tank for collecting waste, and then she arranges for trenches to be dug for placing the pipe in the *leech field* where the waste water drains. You'll need to have a septic company inspect this system before work can continue.

Every septic tank worth its salt has an access port to facilitate cleaning. Make sure the opening to pump out the tank is located in a convenient location and is clearly marked so you can find it later. You may want to put a stake by it so it doesn't get covered with dirt when excavation takes place.

Bringing in electricity

Assuming your property is on the public electric grid system, you'll need to tie into it to get electricity into your home. The power either comes in underground or from a pole. Your power company generally gives you between 25 and 60 feet of wire — from the power pole to your home — at no charge; anything after that costs extra. For example, during the course of Kevin's custom home project in Northern California, the power company charged $15 to $25 per foot of extra wire depending on the difficulty and labor. However, the company also offered a credit of $1,053, allowing for a distance of 53 to 75 feet for free.

Needless to say, even considering the free wiring you'll automatically get credit for, if you have a big lot, running power to your home can get costly. You may want to check with your contractor about having his laborers do the trenching work to see if it saves any cost. If you do, don't forget that all the work must meet the electrical code.

Your contractor sets up a temporary post with a circuit breaker and a meter so the utility company can start measuring your usage immediately. Your building inspector needs to inspect this temporary meter before the electric company will connect the power.

Piping in gas

If you're connecting to a public gas system, then your contractor will arrange for the pipes to be trenched that will run from the gas line under your street into your home. Details for this routing need to be noted on your site plan.

If you're running propane gas, your supply company will usually handle the entire installation process, including delivering and securing of your propane storage tank. However, you don't need to turn on either the gas or propane prior to the installation of appliances later in the project. You and your contractor need to decide together the best time to install this system. We discuss the topic of installing your rough gas piping in Chapter 13.

Spot-check — Utilities

With the utilities in place, you're ready to start the building process in earnest. Make sure you spot-check connections before you move on to the next step in the process. You need a level and a tape measure.

❑ Water pipes don't leak and connections are secure. Spigot is clearly flagged.

❑ Sewer pipes are well buried and marked at the connecting point. The inspector has signed off on the septic system.

❑ Power poles are secure with no exposed wires.

❑ Gas pipes are buried, and connections are clearly marked.

Take time to talk to your contractor about any issues or discrepancies and discuss possible solutions.

Pouring Your Foundation

Because so much depends on your home's foundation, you may argue that the foundation is the most important part of the home-construction process. Get it right, and your home will provide you and your family with decades — perhaps even centuries — of service. Get it wrong, and you may find yourself living in your worst nightmare.

Your foundation is the base of the house. It's responsible for supporting all the weight of the walls, the roof, and, of course, the people who live in it. Your architect and engineers will create specific foundation plans and calculations for your subs in the working drawings discussed in Chapter 6. These plans let them know where everything goes and how much dirt needs to be moved to make room for the foundation.

Your engineer plays a big part in figuring out which type of foundation is appropriate for the kind of home you've decided to build and the characteristics of the lot on which you plan to build it. You can read more about the engineering process in Chapter 6. You can choose from a variety of foundation types, and every local government has different requirements. Although we explain the basic types in this chapter, you need to discuss any variations with your contractor so that you understand what needs to be built on your property. These decisions will likely be made with your engineer earlier during the planning stages discussed in Chapter 5.

Regardless of the type of foundation, scheduling is critical. Some cities and counties don't allow any grading during the winter, so coordinating the schedule with your contractor is important. Also, weather can play a big role in delays during the foundation construction process (which will lead to delays in your overall schedule). Excessive rain, for example, can cause erosion problems, and concrete — which plays a large role in most foundation work — requires six to eight hours of dry weather before being poured.

Excavating the site

Before your contractor and the specific sub pour the foundation, they'll need to first dig out all the dirt from where the house will be. (Check out the "Clearing and grading" section, earlier in this chapter, for more information about the actual process.) During the excavating stage, your lot may look like a playground filled with bulldozers, diggers, and trucks, ready to make pits, holes, and trenches.

The first step in excavation is removing the topsoil from the foundation area. Then, the sub digs holes for the *footings* (reinforced sections at the base of the foundation) if required. These holes will be several feet deep in cold climates so that they're below the *frost line* (the maximum depth that the ground will freeze in winter). The type of foundation you install determines how much dirt to remove. If you're building a basement, for example, you may have to dig down as much as 8 feet, which is a lot of dirt that will soon need a new home. Anyone for some free dirt? (The excavator can pile some of the dirt in a corner of your lot so you can use it during landscaping.)

Installing the foundation

In this section we detail the installation process for three very common foundation structures. These basic descriptions should help you understand how your house is supported and attached to the ground.

The perimeter footing foundation

The most common variety of foundation, built on relatively flat or gently sloped soil, is the perimeter footing foundation. It's called the perimeter footing foundation because it comprises concrete footings that go around the house's outer perimeter.

After the concrete is completely *cured* or hardened (a process that may take several days), the sub removes the forms. In the case of the perimeter footing foundation, walls are built primarily on the house's perimeter; however the engineer may have specified certain places where an interior wall needs to be built to support extra weight. Additionally, the contractor may install intermediate support bearing posts to help carry the weight of long stretches of flooring. These posts are set in concrete in the middle of rooms.

Other than California — which has to worry about earthquakes — many states allow the construction of cinderblock foundation walls around the perimeter. These types of walls are hollow cement blocks that can be stacked and filled with rebar and cement. Your architect and engineer can determine the best approach for your particular home.

The concrete slab foundation

If you're building a slab foundation, then it will be made entirely of concrete. Slab foundations are less popular for custom homes because they don't allow for any sort of a basement, and they make the first floor feel very rigid. However, on the plus side, slab foundations are fairly cost effective compared with other options. If your house is small and relatively inexpensive, your architect may have chosen a slab foundation. Regardless of what kind of foundation your house has, your garage is likely to have a slab foundation due to its ruggedness and relatively low cost.

When the concrete is allowed to set, make sure you put in a handprint or two with names to record the moment for posterity.

Pier-and-grade beam foundation

A very popular form of foundation today is the pier-and-grade beam foundation. This foundation is versatile because it requires little or no excavation. It's often used for hillside application because it secures the house deep into bedrock, preventing potential slide and settling issues. It's also popular in seismically active places such as California due to its earthquake resistance.

Some contractors like to use pier-and-grade beam foundations, even on flat lots, because they can require less concrete and less excavation, resulting in lower costs.

Spot-check — Foundations

Now that the foundation is finished, you need to have a clear picture of how the house will sit on the lot. Use this spot-check to be sure you're ready to move on. You need work gloves, a tape measure, and a level.

❑ All concrete is hard and secure, and there are relatively few cracks. No rebar is showing.

❑ The foundation hardware is securely in place and not bent or broken.

❑ Crawlspaces and utility connections have access points.

❑ All protruding utility pipes are clearly marked and have protective foam collars.

❑ All corners are square or at proper angles per the plans.

Alert your contractor about any issues or discrepancies and then figure out when and how to fix them.

Chapter 13

Framing and Rough: So Much Goes Behind Those Walls!

*I*f you think of a home as a body (okay, bear with us for a moment), then the framing is the skeleton, and the *rough systems* — your home's electrical and technology wiring, plumbing, heating, and air conditioning — represent the guts. After you have the foundation finished (we touch on foundations in Chapter 12), you need to give your creation strong bones and the systems to move air, water, and energy. And that's exactly what this chapter is all about.

This chapter provides you with the framing and rough systems basics that you need in order to have meaningful conversations with your contractor and subs and to be able to follow up on their progress. Throughout the chapter, you find short lists of spot-checks you can make during the construction process to be sure that your contractor and subs are sticking to the plan (and sticking to your budget!).

Things to Consider Before Framing and Rough Work Begin

Believe us, you have a lot to consider before the framing work actually begins. You need to think about everything, including what materials to use, communication schedules, and what questions you want to ask or points you want to clarify with your contractor. Don't worry; we help you with this part of the heavy lifting.

But, before anyone picks up a hammer or nails, you and your contractor need to review the *plans* (you know, those big pieces of paper with blue lines all over) for marked flooring, framing, roofing, electrical, and plumbing. (You can find more information on blueprints and plans in Chapter 5.) The detailed engineered plans, known as *working drawings,* have specific instructions for framing, subflooring, rough electrical, and rough plumbing. A precursory review insures that everyone is on the same page about the tasks ahead.

Reviewing the specific drawings with your contractor and subs at least once a week is a good idea. This review helps manage your expectations and keep you informed along the way. Checking out the drawings gives you a perfect exercise for the weekly planning meeting we recommend in Chapter 11.

Questions to ask your contractor

Understanding the stages in the building process can give you comfort and piece of mind as many different people spend lots of your money throughout the build. Because most people have the least understanding about the framing and rough systems stages, you need to ask your contractor and subs the following questions:

- ✔ What is each sub responsible for?
- ✔ What are you (the homeowner) responsible for?
- ✔ Which subs are involved and when are they involved?
- ✔ What is the proper order of tasks?
- ✔ Who's responsible for managing the coordination of tasks?
- ✔ What are my responsibilities with inspections and when will they occur?
- ✔ How does the contractor plan to keep you (the homeowner) apprised of the schedule, costs, materials, and so on?

You'll continue negotiations on materials and labor throughout your project. Even though the contractor gets bids early on in the process, prices may change. After you have the subs' estimates for materials and labor, always discuss with your contractor whether you can negotiate even better deals by paying the subs cash or on a weekly schedule. If you have structured your finances for maximum liquidity, as we discuss in Chapter 7, you should be able to customize the project payment plan to give yourself the maximum buying advantage (and the best prices).

Establishing a schedule

When building your own home, the schedule is *everything.* When your schedule starts dragging, not only do you become frustrated with an apparent lack of progress, but also the overall completion of your home may become delayed — sometimes by weeks or even months. And, guess what? Big delays can often cost you big money.

So, you have to stay on top of the building schedule. Discuss with your contractor how long each of the subs should take to finish their part with the framing and rough work. Times can vary greatly, depending on the size and complexity of the house, availability of materials, weather, and so forth — from a few days to a few months, or even more. Having a schedule doesn't guarantee everyone can finish on time, but it does allow you to track progress and make adjustments, if you need to. For example, because you can bill most labor on an hourly basis, monitoring the schedule (and keeping your contractor and subs on it) can help prevent the additional costs of paying electricians to sit and wait for the framers to finish. See Chapter 11 for more on scheduling with your contractor.

Trusses and I-beams: Yes or no?

Before the framing phase begins, you still need to discuss with your contractor whether or not to use engineered lumber such as floor trusses and I-joists. *Floor trusses* and *I-joists* are prefabricated wood structures commonly used for framing subfloors and roofs. Although floor trusses and I-joists often cost more than the materials that you buy for custom framing, because they're made by combining different kinds of wood and glue — in a process called *composite construction* — they're often stronger and the quality is more consistently made (and therefore worth the extra expense). Your engineer may designate the use of these materials in your plans (see Chapter 6) but you may have some room for the contractor's interpretation and choice of specific materials.

Trusses and I-joists are useful for running long distances when you have no underlying support, such as in an open cathedral ceiling, for example. In addition, when you factor in all the material waste and labor associated with custom framing, trusses or I-joists may end up being more economical in the long run. Many builders are using another composite material called laminated veneer lumber beams (commonly referred to by brand names such as *Microlam* and *Paralam*) extensively in their building process because of its versatility, uniformity, and cost benefits. Lastly, engineered composite plywoods are used in many cases to minimize the creaking sounds that keep everyone up late at night. This choice may or may not be more expensive than using standard plywood, but it does help the house retain a modern feel as it lasts through time. Common composite plywoods include *waferboard* or *oriented strand board*. Our advice? Discuss the pros and cons of using engineered materials with your contractor before the framing process begins.

Looking at What's Involved in Framing Your House

Someone has come up with a way of framing a house out of every material on earth. Grass, trees, stone, reeds, brick, adobe clay, steel, and concrete — someone, somewhere, has used these materials to build homes. And contractors are constantly reinventing old methods or using innovative new ideas for home construction. Modified old methods like *rammed earth,* where the contractor compresses dirt between wood, stacked hay bales, or even recycled automobile tires and beer bottles, are all gaining attention as structural materials today. This section walks you through framing methods step by step for the most common materials and gives you a spot-check to discuss with your contractor.

Understanding the framing process

By now, you and your contractor are pretty clear on the method of framing to be used in your home. We extensively discuss choices of framing options, such as steel, timber frame, and log in Chapter 4. Regardless of which framing technique you or your contractor decides to use for your home, if your home is being framed on-site, the process order is the same. In this example, we use the most common framing method that most people refer to as a *stick-built* home.

Most people think that framing refers only to putting up walls, and the term conjures up images of an Amish barnraising. Framing actually refers to the entire structural skeleton of a house, not just its walls. The framing process consists of the following phases, generally done in this order:

- ✔ **Subfloor:** The base floor is built below your carpeting or finished flooring and on top of the foundation. A subfloor isn't always required for a slab foundation but is necessary if the house has a basement.

- ✔ **Load-bearing walls:** These walls are added to support weight on top of the subfloor.

- ✔ **Interior walls:** These walls are erected, along with spaces for doors, windows, fireplaces, and architectural structures, such as dormers and soffits. (See our architectural amenities list in Chapter 5 for definitions of these structures.)

- ✔ **Second story:** The framing sub repeats the previous three bullets to create an additional story if required.

- ✔ **Roof framing:** The roof's frame is put into place.

The incredible, shrinking board?

The true size of the 2 x 4 that you find in any hardware or home supply store actually isn't 2 inches x 4 inches. In reality, a 2 x 4 is more like 1.5 inches x 3.5 inches. The term 2 x 4 represents the size of the wood before it's processed for use. Because wood goes through many finishing and drying processes before it's sold, the board has shrunk a lot by the time you can finally get your hands on it for use in construction.

Starting at the bottom with subflooring

Gone are the days when people had little choice but to sleep on the cold dirt, with nothing more than a pile of leaves to act as a pillow. Nowadays, you have floors and beds to keep mice and spiders from building nests in your hair as you sleep at night. After finishing the foundation, the first step in the framing process is putting in a *subfloor,* which serves as a platform for building walls and creates a base to attach flooring. Putting the subfloor in is mostly a wood-based process, and the framer, carpenter, or general contractor generally build it.

Subfloors aren't always necessary, however. If you're building your house on a concrete slab foundation, for example, you don't need a subfloor because your entire house is supported directly by the slab. The following list gives you the instances when you need a subfloor:

- ✔ If you build the house on raised or pier-grade beam foundations (see Chapter 12 for details)
- ✔ For the upper floors of multistory houses
- ✔ As a base floor underneath hardwood floors (even if you've built the house on a concrete slab foundation)

Here is a brief list of parts in the anatomy of the subfloor:

- ✔ **Girders:** A long heavy piece of wood that spans the floor and provides support.
- ✔ **Joists:** Long narrow boards set in place on their edge resting on sills and girders. The outer joists are the *header* and *stringer joists,* and the inside joists are called *lap joists.*
- ✔ **Seal:** A metal strip or layer of sealing material called *caulk* designed to protect the wood framing from moisture and bugs. It sits between the footing and the sill.
- ✔ **Sill** or **sill plate:** A piece of pressure-treated wood anchored to the foundation.
- ✔ **Subfloor:** Plywood lies flat over the joists usually at a perpendicular angle to direction of the floor joists.

To avoid squeaky floors, be sure that your flooring sub glues down the plywood when building the subfloor, and that she uses a *ring shank* nail to secure the plywood. The nail has screw threads that hold tight.

And, while you're at it, make sure that your sub is using flooring plywood graded CD exterior or CDX, which is more durable to weather during construction because manufacturers make it with waterproof glue. You can identify CDX plywood by inspecting the identification stamp found on each sheet.

Be sure that the subfloor has access for all the heating, ventilation, and air conditioning (HVAC) ducting and plumbing work that you need to complete under any flooring before it's sealed up! Work with your contractor to stage the process so that you don't have to tear out work that the builders have already completed. Going back to do work again can cost money and delay the entire project.

Sharing the load — Load-bearing walls

After the subfloor is in place, the next step is to put up the walls. Framed walls have a very simple structure. Whether you're building with 2 x 4 or 2 x 6 studs based on your plans, you use the same basic elements.

A bottom plank, called a *sole plate,* is the base for a series of studs (usually 92⅝ inches in height) set 16 inches apart. Contractors take most measurements for spacing from the center of any given stud to the center of the next one. They usually note the measurements as "16 inches O.C." (or On Center). Sometimes to save material in nonload-bearing walls, they may place studs 24 inches O.C. A piece of lumber matching the sole plate caps these studs and is called a *top plate.* A *tie plate* above the top plates ties the walls together at the junctions. The entire structure measures 8 feet 1⅛ inches from top to bottom for a standard 8-foot ceiling. Note that custom homes vary greatly and your plans may call for taller ceilings. Exterior walls are generally framed first. They will be braced in place and then attached to each other moving around the house's perimeter. The framing sub creates spaces for windows and doorways that don't detract from the structural integrity of the wall in the framing. The sub does this job with a header. A *header* is a large, strong piece of lumber designed to distribute the wall's weight away from the opening for the door or window. *Trimmers* are side supports that transfer the weight from the header to the sole plate. *Cripples* are small pieces of wood that secure narrow gaps and make up wall height.

If assembling a log home, your contractor will be given specific instructions on how to stack the logs. If the logs are milled in a factory, they'll likely interlock with a sealer inserted in between, but if the logs are handcrafted, then extra shaping may take place on-site and an insulating gooey material called *chinking* will need to be applied between the logs. In either case, the logs will be bolted together as they're stacked.

Other kit homes, such as panelized, cedar, or timber frame, have their own individual methods of assembly, but in most cases, the materials have been precut and possibly preassembled at one time before being disassembled and shipped to your site, which means that doors, windows, and other architectural elements have been placed and accounted for before framing starts.

The plumbing sub may need to install large plumbing fixtures, such as bathtubs and Jacuzzis, before you finish the framing so they can fit them into their locations. If building materials — especially _expensive_ building materials — in your community have a habit of sprouting legs and walking away, you can chain them to the studs to prevent theft.

Framing the second story

Currently, contractors use two different methods of framing for multistory residences:

- **Platform framing** (also known as _stick framing_) is the most common framing method and allows for multiple levels with unique architectural choices.

  In platform framing, the framing sub builds the first walls attached to the subfloor. The ceiling _joists,_ or supports, then serve as floor supports or joists for the next story. After installing a new subfloor, the framing sub builds the second story walls on the subfloor platform, just like he did on the story below. The sub needs to construct stairways to allow workers to access upper build sites at minimal risk.

- **Balloon framing** uses continuous studs running from the sole plate at the base of the house up to the roof.

  This method allows for the transfer of the weight of the roof and second story directly to the foundation. Because this method requires a uniform outer shell of the house, balloon framing can be architecturally restrictive and is used much less than platform framing.

Finishing the architecture

The next step is to frame out all those wonderful architectural pieces you had designed for the house. If you worked with an architect, all the time and money you spent will pay off for you here as long as you stick to the plan. (See Chapter 5 for the discussion on what an architect can do for you.) If you chose to work without an architect, you'll now find out if you made the right choice. During this stage is your last chance to make any structural changes to the shape of the house, and expect additional costs even if you change at this stage. If you must make a change, we tell you how to deal with change orders in Chapter 11.

Next, the framing sub builds the interior walls, and your home's interior architecture begins to take shape. Your sub frames your fireplace(s) at this time, too. Your subcontractor may need to construct brick and stone fireplaces before closing in the floors and walls.

Dormers, decks, and soffits are built out in an almost sculpturelike fashion, using the same spacing and support structures as the walls and subfloors. All external structures, such as attached decks, require pressure-treated wood and galvanized nails and hardware to make them weather resistant. Scientists are continually creating new chemical advances to prevent rust and deterioration, so talk to your contractor about using the latest and greatest materials.

Framing the roof

The roofing subcontractor installs the roof using either individual rafters or prebuilt trusses or I-beams (see the section "Trusses and I-beams: Yes or no?" earlier in this chapter). Because you can find multiple roof styles and types of roofing materials, the structural design varies according to the span and weight requirements of the different materials. For example, clay tile and slate are heavy, requiring stronger framing support than a lighter asphalt shingle.

While the house is being framed, you can spend time with your contractor examining the roofing plan in your working drawings. (For more about working drawings, see Chapter 6.) Your contractor uses those plans to determine the materials, labor, and amount of time necessary to finish the roof framing.

Spot-check — Framing

So, how is your contractor doing? Go through this handy spot-check to discuss whether you and your contractor are ready to move ahead. You need a tape measure, level, T-square, and flashlight.

❑ All measurements and materials match those called for in the working drawings within a quarter inch.

❑ All angles with T-squares and levels are straight and plumb.

❑ Floor joists don't interfere with plumbing and drain systems.

❑ No loose, damaged, weak, or unreinforced boards are in place.

❑ All lumber in contact with concrete is pressure treated (ask your contractor about the lumber she used).

❑ No squeaks in the subfloor or on the stairs, and measurements are uniform.

Discuss any issues or discrepancies with your contractor and determine when and how to fix them. Before you move on to the next phase, you may need the building inspector to come out and inspect the framing. Communicate with your contractor as to the time and date for this inspection. (We discuss dealing with inspectors in Chapter 6.)

Installing the Rough Systems with Ease

Now that the framing is done, it's time to add the mechanical elements that make a house livable. Where would you be without water, hot (or cool) air, and power? The working drawings for your home have specific plans for plumbing, HVAC, and electrical systems. Another meeting with your contractor can go a long way in clarifying the schedule as well as the materials used by the subs.

You and your contractor need to check your lighting and plumbing fixture orders while going through the process of roughing out your plumbing and electrical systems. Many suppliers don't store large quantities of expensive fixtures, and you may be unpleasantly surprised to find out that your pieces are on back order on the day you want them delivered. Checking now can save you a lot of wasted time (and money!).

When the construction crew is done with the rough work, pull out your camera and take extensive photographs of your home's rough systems. You may also consider using a video camera to get more perspective. By documenting your home's inner workings, you can locate important structures long after you've finished the walls. This documentation can be helpful for repairs and renovations decades from now.

Rough plumbing

Plumbing — all those steel, copper, and plastic pipes that route water into, through, and out of your home — is actually composed of three separate but interrelated systems: the water supply, the sewer system, and the vent system. The *supply* brings water to different places in your home, and the *sewage* system takes it away. The *vent* system is necessary so that the water can move in a steady flow without bubbles or spurts.

At this point, your contractor marks the location of all plumbing fixtures on your home's framing, according to the working drawings. You probably want to discuss and mark where plumbing should *not* go at this time because of planned light fixtures, built-in cabinets, or for sound insulation reasons (nothing's much worse than having a regularly flushed toilet located only two or three feet away from your head when you're trying to get a little extra sleep on a Saturday morning!).

You need to walk through the house with your contractor now to discuss material choices, fixture choices, and any special needs or conflicts that have developed since you've framed the house.

For reference, here's a brief checklist of common plumbing fixtures that you need for your home:

- ✔ Bathtubs and faucets
- ✔ Exterior water spigots
- ✔ Garbage disposal
- ✔ Refrigerator or icemaker
- ✔ Sauna or steamroom (mandatory for our Finnish readers!)
- ✔ Septic tank, if applicable
- ✔ Shower fixtures
- ✔ Sinks and faucets — bathroom, kitchen, utility, wet bar
- ✔ Toilets, toilet seats, and bidets
- ✔ Water heater
- ✔ Water softener or conditioner

Opening the flood gates — Installing the water supply

In most cases, you can have water piped into your home from a public utility (the water company) or from a well on your property (we cover the subject of wells in great depth in Chapter 12). Chances are, your contractor has already installed a meter and spigot for his use during construction, so your house gets hooked up now. For the next step, your plumbing sub installs the infrastructure necessary to distribute this water supply from its primary source to every place within your home where you need it. And, believe us, you're going to need water in a *lot* of places in your home.

During installation, the plumbing subcontractor lays both hot and cold water lines in the framed walls and crawl spaces. To help keep it safe from stray nails or other pointy objects — both during and after the construction process — pipe should be run through drilled holes in the studs rather than notches, which by nature put the pipe up against the wallboard. Your plumbing sub installs pipe supports and further protects the pipe from being pierced by nails by using protective 1½ inch metal strips with holes called *FHA straps* or *nail guards*.

Although galvanized steel pipe was once the material of choice for plumbing homes, most people today choose copper tubing. Why? Copper tubing is durable, readily available, easy to work with, and relatively compact. On the other hand, even though copper tubing can cost twice as much as steel pipe, the labor savings have made steel virtually obsolete.

To plumb or not to plumb?

Ever wonder where the word plumber came from? It turns out that *plumber* — someone who works with water pipes — is derived from the Latin word for lead, *plumbum*. In ancient Rome, they used lead water pipes, and that's where the words *plumber* and *plumbing* come from. Even today, plumbers routinely use lead solder to join together copper water pipes. Coincidentally, the word *plumb* — meaning to make something (such as a framed wall) exactly straight and vertical, also derives from the Latin *plumbum*. In this case, however, plumb refers to the lead weight at the end of a string (a *plumb bob*) that you use to test the straightness of a wall or other structure.

Because of the high cost of copper tubing, builders have been searching for a less expensive substitute. Plastics, such as chlorinated poly vinyl chloride (CPVC), have gained some measure of popularity because they're easy to use and cost a lot less. Some people, however, believe plastic piping leaves an aftertaste in the water. A relatively new plastic, called polybutylene, has gained some popularity due to its flexibility. It allows for less joints and connections, reducing the cost of installation, even though the material may cost more on a piece-by-piece basis. It's virtually freeze-proof, however, which can be a big plus in cold-climate areas. Plastic plumbing isn't yet approved in all cities, so discuss your specific needs and requirements with your contractor and plumber.

Generally, before you turn on your water supply, the plumbing sub conducts a test to check for leaks or blockages in your water pipes, first using air pressure and then using water. This test checks the time and water volume it takes to fill the system. The contractor can rework any leaks or blockages before connecting the system to the water supply.

A home isn't complete without a water heater, and you need to install it in a place where leakage or failure won't likely destroy interior flooring — maybe a closet on the exterior of your home or in the garage. Some people put it in an unfinished basement if one exists, but placing it there doesn't provide easy cleanup if it were to burst (and water heaters do burst, from time to time). At the very least, be sure that you've installed a protective pan underneath the water heater and that you use adequate bracing to support it — particularly in earthquake-prone areas.

Getting down and dirty — The sewer system

Gravity powers your sewer and waste systems, so your lowest sink, toilet, and drain must be higher than the sewer line that leaves your home. If they're lower than the sewer line leaving your home, you're going to be in a major world of hurt soon. The sewer line must have at least a ¼-inch drop per foot in order to operate properly. When properly installed, your sewer line efficiently delivers waste from your home to the city hookup (where it eventually find its way to a sewage treatment plant) or possibly to a septic tank on your property. We cover both of these possibilities in Chapter 12.

Your primary choices for sewer materials are cast iron or poly vinyl chloride (PVC) pipe. Although PVC is cheaper and easier to work with, it has a tendency to be noisy when water flows through it. If you decide to use PVC, take care to position the pipe in walls isolated from bedrooms and media rooms where the sound could be annoying. The drainpipes have occasional openings to the outside of the house, called *cleanouts*. These openings have removable caps so that you can snake out difficult clogs and roots in the future.

If building a basement, discuss the placement of your home's main sewer line with your contractor. Make sure the sewer line doesn't intrude on the living space, or your carpenter may have to frame a post, breaking up the useable area. Your plumber also cuts holes in the framing for all the drain lines that meet with the sewer system. Make sure the plumbers don't cut any I-joists when installing bathtub and toilet drains. They're notorious for doing this!

Getting a breath of fresh air — The (wet) vent system

The *vent system* — sometimes more completely known as the vertical wet venting system — allows your home's plumbing system to breathe. Without this venting system, the plumbing would be ineffective due to problems like siphoning and backpressure, which can stop the water flow (and believe us, you do *not* want to stop the water flow in your home!). Think of this vent system like the way water burbles slowly out of an open bottle when you turn it upside down. If you look at the outside of a house, you notice narrow vent pipes extending from the roof. All the drains in the home attach to one of these vents, allowing air (and all those smelly gases) to escape.

In addition to installing the water systems, either your plumber or your HVAC sub installs pipes for the natural gas or propane system if you have gas service to your property. The process is similar to the water-supply plumbing. Additionally, the contractor installs the pipes for radiant heating, which we discuss in Chapter 5, at this point.

Make sure that pipes remain uncovered until the county or city building inspector has signed off on the plumbing. You need to be available to walk through your home with the inspector so that you can be aware of any issues as they arise.

Spot-check — Plumbing

To get a good look at your plumbing, discuss the plumbing with your contractor and go over it all with her. You need a flashlight for this spot-check to make sure that you can see in those dark parts of your home where plumbing just naturally seems to prefer to live.

❑ Vents are in hidden areas.

❑ No moisture has collected at joints.

❑ Pipes inside walls are securely fastened to prevent knocking.

❑ Water heater is secure and has a pan underneath it.

❑ Cleanouts are accessible but not going to affect the look of your interior.

Make sure you talk over any issues or discrepancies with your contractor and discuss possible solutions.

HVAC

Your house may require multiple HVAC units for efficiency. You need to make these choices in the design phase (see Chapter 5). When you install the units, whether in the attic, crawlspace, or closet, be sure to discuss noise-reduction options, such as vibration absorbers and insulation, with your HVAC subs.

Installing the mighty ducts

The ductwork for your HVAC creates passageways for air delivery and return throughout your home. The ducts are usually made of two materials, cut and fit at the build site. The two materials are

- **Insulated fiberglass:** Fiberglass ducting is becoming more popular because you can cut it with a knife, and it has greater insulation, making for lower installation cost and quieter airflow.

- **Sheet metal:** Sheet metal ducting can develop noise over time because of temperature changes causing the metal to expand and contract. Sheet metal is becoming less popular because it's more expensive and harder to work with.

The openings in the wall for air supply and air return are called *registers*. Correct placement of registers is critical to the efficiency of your HVAC system. By putting registers near doors and windows, the conditioned air compensates for the heat or cold loss generated by these openings.

Each room needs to have at least one supply duct and one return duct. For heat, which rises to the top of the room, you need to place the supply low and the return high. The opposite is true for the cold air system. Discuss convenience and efficiency of register placement and quantity with your contractor and the HVAC sub.

In addition to ducting for the climate system, you need to install *power ventilation ducts* for bathroom and kitchen exhaust fans. These ducts require electricity as well as ventilation to the outside. Discuss these ducts with your contractor and mark them for placement.

Installing your thermostats

Although your contractor won't permanently install HVAC thermostats until you finish putting up the drywall, you must decide with your contractor on the best locations. (The location of your thermostats probably won't be designated on the working papers. You need to revisit their locations with your contractor during the rough work stage.) Here are some factors to consider:

✔ Don't place the thermostat within 6 feet of a register or on the opposite wall.

✔ Interior walls work better because outside walls can absorb outside temperature.

✔ Don't place thermostats in rooms with a fireplace.

✔ Hallways make for great thermostat placement.

If building in cold weather, consider getting your heating signed off by the inspector right away. Running the heat during the building process may make for more productive subs, allowing them to spend less time warming up in front of space heaters that you're paying for. The more temperate climate your heating system creates also helps with the drying process of the drywall and paint, covered in Chapter 14. You can also ask your contractor about installing a temporary furnace if you're in a cold area and the HVAC system is taking a long time to install.

Spot-check — HVAC system

Wonder if your contractor's breezing through the HVAC installation process? The following spot-check gives you everything you need to check out before you give him the green light to move ahead. You need a flashlight, screwdriver, and Crescent wrench for this check.

❑ Get your ducts in a row. Make sure that they don't have any punctures, cracks, dings, crimps, or nail holes.

❑ The HVAC is securely fastened.

❑ The HVAC unit's working parts all look new under the panel. Corrosion can happen in a unit that sits outside too long.

❑ The registers are all in the appropriate places with clear airflow.

❑ The ducts are fastened and insulated in crawl spaces, walls, and the attic.

Take up any issues or discrepancies with your contractor and subcontractors and determine when and how to fix them.

Rough electrical

Before the electrical work starts, your electrical sub reviews the electrical plan in the working drawings with you and your contractor to see if you need to make changes. Your sub is a licensed electrician, and it's her job to make sure that she gets all electrical choices installed to code.

Building departments most often use the National Electrical Code (or NEC, for short). The NEC is available through a number of different sources on the Internet for about $75; one good source is the National Fire Protection Association Web site (www.nfpa.org). Simply pay your money, and enjoy the light reading.

You may want to walk the house with your contractor and electrician now to mark where all your switches and receptacles (outlets) will go. Although your electrical plan addresses the minimum number of switches and receptacles required by code, you need to decide where, specifically, you want to have them placed on the walls. You also may decide to have more switches and receptacles installed for your convenience.

Some additions you may want to consider include

- Additional outlets around kitchen counters and breakfast areas
- Dimmer switches for more romantic lighting
- Extra outlets in the home theater areas, hobby rooms, and home office
- Switched exterior outlets at the patio and in the soffit for holiday lighting
- Switched outlets in the attic and basement
- Switches at the top and bottom for stairway lighting
- Three-way switching in bedrooms near headboards so you don't have to get up to turn off the lights

Wiring your home for electric power

Your local building code specifies the gauge and type of wiring that you have to use. You can discuss with your contractor and electrician which materials give you the best protection regarding age, corrosion, and fire. For example, *Romex,* a brand name for a type of plastic insulated wire, is often used. (Sometimes called nonmetallic sheath, the formal name is NM.) Some contractors also install metal *conduit* (pipe) as further protection against damage.

Your electrical sub runs cable or conduit throughout the house in different loops, or *circuits.* These circuits distribute the electrical load and tie back into protective circuit breakers at the main panel. Your electrical sub can install subpanels — which further divide key circuits — for your convenience. He needs to mark each of these circuits on the service panel so that you can identify which breaker goes to which outlet. This attention to detail saves time when you're trying to make repairs in the future. You want your sub to mount panels in unobtrusive but convenient locations, such as in the kitchen, or by your garage door so that you can easily locate them if and when a breaker trips. The main service panel has a safety switch and a wire connecting it to *ground,* which is a water pipe or metal stake going into the ground.

Many people typically take for granted how much of life depends on electricity. To make sure that you haven't neglected an electrical must-have, use this checklist of items that require power in specific locations:

✔ Attic fans	✔ HVAC
✔ Bathroom heater	✔ Interior lighting
✔ Bath vent fans	✔ Kitchen fan
✔ Ceiling fans	✔ Light switches: One- and two-way
✔ Dishwasher	✔ Microwave
✔ Doorbell	✔ Sprinkler system
✔ Exterior lighting	✔ Stove and oven
✔ Garage door opener	✔ Sump pump
✔ Garbage disposal	✔ Washer and dryer

Wiring your home for technology

In addition to wiring your home for electric, you want to address wiring your home for technology at the same time. Because technology is constantly changing, wiring your house in a way that can last 20 years or more can be very difficult. Until recently, you absolutely needed to install CAT-5 computer network wiring if you wanted Internet access throughout your home. Today, however, wireless systems are all the rage, and CAT-5 may soon be obsolete.

Allow for future innovations, regardless of the onward march of technology, by running open conduit throughout the house with an outlet in every room. That way, you can pull wires anytime you need them. Doing so requires a *pull-wire system.* A firm but flexible wire is fed through the conduit and remains there until you decide to pull cable. By attaching the cable to the end of the wire, you can pull the cable through the conduit. Attaching and pulling an additional pullwire through before you seal everything up is best so you can add new cable in the future. This open conduit may add some additional upfront costs, but it saves you money five years down the road when you need to install the latest and greatest technogadget that requires different wiring than currently exists in the house.

Keep these additional systems in mind when wiring:

✔ **Computer:** Wireless is gaining popularity, but CAT-5 wiring and Ethernet jacks provide added security.

✔ **Security system:** Common practice is to wire every door and window.

✔ **Smoke alarm:** Hardwiring systems to your electrical system prevents constant battery changes.

✔ **Sound system:** Pipe that music everywhere without tripping over speaker wires.

✔ **Telephone:** You may want to put a jack in every room.

✔ **TV/Cable:** Like phone jacks, you may want to put a coaxial outlet in every room.

✔ **Vacuum:** Central vacuum systems are popular in big houses, requiring special pipes and outlets in every room.

Spot-check — Rough electrical

If you want to see whether your contractor has the right electrical stuff, see how his work fits in the following list. You need a flashlight and tape measure.

❑ All wire splices are protected and put in junction boxes (also called *J boxes*).

❑ All circuits are marked in the service panel board, leaving no loose or unidentified wires.

❑ You have approved Romex or NM, wiring, or conduit, per your plans.

❑ You have offset outlets and switchboxes to allow for drywall and base molding.

❑ All electrical boxes are adequately braced for ceiling fixtures to hold heavy fixtures, such as ceiling fans and chandeliers.

Tell your contractor about any discrepancies that you notice and determine what needs to be done to fix them.

Sheathing, Flashing, and Insulation

Normally, sheathing, flashing, and insulating are the last stages of buttoning up the house before moving on to the finish work. However, depending on the time of year and the climate in your area, your contractor may choose to sheath the house before installing the electrical, plumbing, and HVAC systems. (For example, if you're building your house in Seattle, then your contractor may want to button up the house quickly to protect it from the constant rain.)

Covering the framing — Sheathing and flashing with your clothes on

Enclosing the framing offers protection from the elements, but leaving it open makes for easier access. Your construction crew encloses the house during the sheathing process. In this process the crew applies a covering material to the outside of the framing.

Your sheathing serves several purposes:

- It adds shear strength to the walls.
- It provides backing for the siding.
- It stabilizes the studs from bending under the load and warping with the weather.

Most sheathing today is asphalt-soaked fiberboard. A composite material commonly referred to by its brand name Built-rite, this fiberboard is the most common and cheapest insulator for sheathing. Other newer and more expensive materials may be available for sheathing. Discuss these options with your contractor as well as the timing of sheathing in your project.

Your contractor uses special nails with protective plates to keep from puncturing the outer skin of the sheathing when attaching to the studs. The crew nails wall sheathing to the stud vertically about every foot. Over the top of the sheathing, a permeable vapor barrier is applied, underneath the siding. The most popular name for the product is Tyvek by Dupont. This barrier keeps moisture out but allows air to pass. Make sure all seams are taped, and windows wrapped with window wrap.

Roof sheathing is generally made of the weatherproof plywood that we discuss in the section "Starting at the bottom with subflooring," earlier this chapter. Your roofing sub needs to secure roof sheathing along the rafters or trusses every 8 inches. When sheathing the roof, the crew needs to allow for whatever type of roof ventilation that you plan to use. The crew may need to leave spaces at the ridges and cut holes for extruding vents and skylights. Any items punched through your roof need to have *flashing,* a metal stripping nailed in place and sealed or caulked to prevent water from leaking into your home.

After you have the roof sheathed, contractors commonly stack the roofing material on it in preparation for the next phase of construction. These stacks of material provide the appropriate weight to the roof sheathing, preventing unnecessary expansion and warping.

Rolling out the insulation

Now that everything is enclosed, your contractor needs to make your house warm and cozy. You and your architect should have already discussed the type of insulation for your house during the design phase discussed in Chapter 5; now it's time to put your plans to work. Your construction crew can easily roll the commonly used blanket insulation out into the walls and ceilings.

If your supplier delivers the insulation to the site early, make sure that you keep it dry and loosely stacked. Wet, damp, or compressed insulation loses its insulative qualities and may become moldy or dirty. Be careful to use work gloves when handling the insulation because the fiberglass strands can irritate your skin.

Walk the house with your contractor and discuss which areas can handle hand-packed insulation and additional soundproofing. Make sure that you have the plumbing and ductwork completely covered and that the insulation doesn't cover any vent holes.

You may want to take this time to have your contractor soundproof those trouble areas. Soundproofing material comes in 4-x-8-foot sheets, and the crew can apply it to kitchens, bathrooms, HVAC areas, home theater rooms, and kids' rooms to provide much-needed peace and quiet throughout the house.

Spot-check — Sheathing, flashing, and insulation

Time to make sure that your house is bundled up just the way you want it. Check your contractor's work, and discuss it with her, using the following list. Has she covered all the points? You need a flashlight and work gloves for this spot-check.

❏ Your contractor has attached sheathing every foot in the middle of the panel, every 6 inches on the edges vertically along studs, or every 8 inches for roof deck.

❏ No nails are loose or extended.

❏ All gaps are taped over a quarter inch.

❏ Skins of sheathing aren't torn, ripped, or broken by nail heads.

❏ All skylights, chimneys, and vent pipes are sealed and flashed.

❏ All wall insulation is securely fastened.

❏ All windows, doors, cracks, and joints have hand-packed insulation, as appropriate.

If you notice any issues or discrepancies, discuss them with your contractor and determine any possible solutions. Before you move on to the next phase, you may need the building inspector to come out and inspect the rough systems. Communicate with your contractor as to the time and date for this inspection. (We discuss dealing with inspectors in Chapter 6.)

Chapter 14

Heading for the Finish: So Much Detail

. . . .

In This Chapter

▶ Finishing the exterior and driveway

▶ Putting in drywall and fireplaces

▶ Installing the doors, windows, trim, and cabinets

▶ Painting and installing fixtures and flooring

. . . .

*I*f you find yourself reading this chapter, then you know that you're in the home stretch — your foundation is complete, and you've framed and roughed the house. The good news? During this stage is where all the beautiful stuff starts to come together and you begin to see the fruits of everyone's labor — including your own. The bad news? These critical areas require craftsmanship and artistry, and most of these steps take extra time and extra money, and require special attention to detail. (Depending upon the complexity of your house, finish work can take up more than 50 percent of your entire time and budget.) The finish work shows your contractor's true mettle.

This chapter is a guide to understanding the finish construction phase of your custom home project. (We address cosmetic finishing like landscaping and hardscaping in Chapter 17.) This chapter can help you set schedules with your contractor for all the different finish subcontractors and give you a clearer idea what finish work is done during each phase. Much of the work that your finish subs will do is meticulous, so keeping up the regular meetings with your contractor as detailed in Chapter 11 is important.

The Icing on the Cake — Exterior Finishing

Enough staring at all that wood and sheathing, it's time to make this house look like a home. Now that the major construction tasks are complete, your contractor and each specific sub can finish the outside walls, the roof, and the driveway, and install lighting to brighten up everything. During this stage is when your home begins to look like that artist rendering you have been staring at for the last year. And, believe us, that's a very good feeling.

Applying wall coverings

Your new home's walls need to be covered. Whether the house is Mediterranean, Tudor, or California ranch style, it needs some sort of outside surface.

You basically have three choices for outside wall coverings:

- ✔ Siding
- ✔ Stone
- ✔ Stucco

The following sections give you information about each choice. Ultimately, your decision will be made based upon your personal taste along with any guidelines imposed by your community. We discuss these design guidelines in Chapter 5.

Siding

Siding is popular in areas with older-style homes, but you'll find it used most anyplace. Your siding sub has four major types of siding to choose from (assuming your architect didn't already specify a particular material):

- ✔ Wood
- ✔ Vinyl
- ✔ Aluminum
- ✔ Cement board

Our advice is to choose wood if aesthetics are your primary concern. Unfortunately, although wood is the most beautiful siding material, it's the most expensive and requires ongoing maintenance in the form of painting, and can suffer damage from sun, rain, insects, and rot. Vinyl and aluminum — which can be purchased in a wide variety of colors — are the more durable types of siding, but they don't look as good as wood and vinyl doesn't protect as efficiently against moisture as other materials.

The latest thing in siding and a favorite with contractors is cement board. It's sometimes referred to as hardi board or Hardie Board, because it's manu- factured by James Hardie. You can check out this siding material at www. hardie.com. This siding is very cost effective. It goes on easily, is relatively inexpensive, and is impervious to insects, rot, fire, and birds. It holds paint up to 15 years because of its thermal stability, it looks like wood, and comes in several styles including a cedar shake. The biggest advantage to this siding is it's fire resistant. Any type of siding is installed relatively quickly because it's premanufactured and uniform.

Stone

If you have chosen a house with a stone or brick exterior, your subcontractor has two choices:

- **Real brick or stone:** Brick or stone is a fairly expensive proposition and requires special craftsmen called *masons* to assemble it correctly. It provides great beauty but can have difficulties decades later if the mortar starts to fail.

- **Masonry veneer:** Masonry veneer is a covering — made to look like stone, brick, or possibly adobe — that is far less expensive to purchase and install than the real thing. Imagine bricks or rocks that are up to 8 inches wide but only 2½ inches thick. (In states like California where earthquakes are prominent, masonry veneer is the only brick home you can have because of seismic requirements.)

Your sub carefully matches the patterns and cover patches so even your nosy Aunt Mabel won't know it's not real brick or stone.

Stucco

In addition to siding and masonry, you can choose one more commonly used exterior finish — stucco. *Stucco* is a thick, concrete-based substance — similar to the mortar applied between bricks when building a wall — applied to your home's exterior. To install, the stucco sub first mounts *wire lath* (think chicken wire) to your house's exterior walls to give the stucco something to stick to and to help hold its shape. The first coat of stucco will be about ¾-inch thick. After the first coat dries sufficiently, the sub applies a *brown coat* (a smoothing layer that may or may not be brown) to smooth out any uneven surfaces. Finally, after the brown coat dries, a final color coat is applied. A good stucco sub can offer a variety of different textures (and even colors, premixed into the coating) to suit your taste — conservative, wild, or something in between.

Finishing an attached deck

If your plans call for building an attached deck, then it needs a finish. Assuming your deck is complete (if not, then get to it!), now is the time to lay in the top of the deck and decide on a finish. (If you need some help with your deck plans, check out *Decks & Patios For Dummies* by Robert Beckstrom and the editors of the National Gardening Association [Wiley].)

Your deck will get a lot of traffic, so make sure you choose a finish that can stand up to heavy wear and tear, as well as to the elements. A wood deck, for example, is beautiful, but it can show the effects of aging and won't last unless it's refinished periodically (exactly how often depends on the quality of the product you use and the amount of wear and tear your deck receives from usage and weather). The best coating for a wooden deck is a penetrating transparent water sealer that allows you to see the wood, but at the same time protects it from the elements.

Many people believe that a concrete deck doesn't require a finish. This belief is a mistake; even concrete can wear and become stained. A sealer can help keep the concrete deck cleaner and looking better, while making it last longer. You may want to consider rubberizing materials if you intend to use the deck for sports play (such as basketball or street hockey). Some homeowners find that using indoor/outdoor carpet can give their concrete decks a warmer, more comfortable feel. You can consider these options after your house is finished and you've moved in.

Up on the roof — Roofing and rain gutters

You and your contractor should have ordered your roofing materials back in the framing stage and hopefully by now they have arrived on-site and are sitting up on your roof, waiting to be installed. If the materials are here, sit down with your contractor and pull out the roofing plan from the working drawings we talk about in Chapter 6. The roof plan serves as a guide for making sure the roof looks exactly as you dreamed and as your architect intended. The roof plan specifies the types of materials as well as the quantities of materials necessary to finish the job. You'll likely hear your contractor and subs refer to the amount of roofing material to be used in terms of "squares." One square of roofing material equals the amount necessary to cover 100 square feet of roof, a 10-foot by 10-foot area. The plans should call for enough material to cover the total square footage of the roof as well as an additional 10 percent for waste because pieces will have to be cut to fit the edges of the roof.

When you and your contractor order the roofing material, having a little too much is far better than not enough. If you run out and order more, the shingles may not match. Different dye lots of shingles may sometimes vary in their appearance — even if they're marked with the same color on the package. Take it from us: Your roof will look better if the shingles match.

The roofing sub and laborers attach the materials to your new roof's *sheathing*, generally, a layer of plywood covered with a layer of tarpaper, and cover the metal stripping (called *flashing*) installed around the chimney and extruding vent pipes (from furnaces, plumbing, ventilation fans, and other sources). If the sheathing and flashing aren't in place yet, then have a look at Chapter 13 for a detailed description of the process.

Although flashing is most commonly used to seal areas around chimneys and vents, your roofing sub may use additional flashing to cover seams and valley sections of the roof in order to provide further protection from moisture. If your roofing sub makes this suggestion, we advise that you agree to it.

Your roofing sub will also finish the *cornice,* which is the area between the roof edge and the exterior wall also sometimes called the *eave,* the part of the roof that hangs over the wall. (Your architect already specified the type of cornice to be installed in the roofing plan.)

Depending upon your local building department requirements, an inspection of the roof may be required after sheathing and flashing. (We talk more about inspectors in Chapter 11.) After the roof has been sheathed, flashed, and inspected, the roofers will put down tarpaper with staples. They'll roll out a protective roofing felt and attach it to the roof for further protection. The next step is installation of the roofing materials, generally, either shingles or tiles.

Selecting and installing your shingles

You probably picked the type of roof for your home during the design process outlined in Chapter 5. If you're still deciding on materials, we break down the benefits of various shingle types in the following list:

- **Asphalt shingles:** These are the least expensive and most common in housing. These thin shingles are made from fiberglass and asphalt. The lifespan of an asphalt roof is generally 15 to 20 years.

- **Clay tile:** Although beautiful, clay tile is a heavy roofing material and requires a reinforced roof to support it. Clay tile is usually terra cotta in color and comes in different shapes. Newer clay tiles are sometimes made from plastic or synthetic materials and are lighter. Tile is very popular in the West because of its fire-retardant properties. Clay tiles can cost 30 times the price of an asphalt roof, but last 50 or more years.

- **Laminated fiberglass shingles:** These shingles contain the same materials as asphalt; however, they're thicker and can be fashioned with different textures to give the roof different looks. They cost twice as much as asphalt shingles, but last 10 to 15 years longer.

- **Sheet metal:** Metal roofing isn't actually a tile but large rolled plates of metal. It's lightweight, fairly easy to install, and comes in a variety of colors. Sheet metal is very popular in snow areas because it allows for snow to slide off the roof more easily. It's ten times the cost of asphalt shingles but lasts two to three times as long.

- **Slate:** Slate roofing is a stone material that is durable and elegant. Like tile, slate is extremely heavy and requires roof reinforcement. Slate is also pricey, costing 30 times that of an asphalt roof; however, it lasts more than 100 years.

✔ **Wood shake shingles:** Some people still prefer wood shake shingles for their custom home projects because they look great. These shingles cost five times the cost of asphalt shingles but have the same life span. Wood shingles are much less fire retardant than the other shingle types and have been banned in high fire danger areas such as Southern California.

To install shingles, your roofing sub and laborers lay a 9-inch-wide starter strip of roofing material across the eave or edge of the roof. In colder, harsher climates, at least one or two rows of ice and water shield are required on the edge of the roof. Then — just as in the case of tile roofs — the sub and his laborers layer the shingles in overlapping rows from the bottom of the roof to the top and fasten them into place with galvanized roofing nails. Sheet metal roofs will be rolled out or have premanufactured interlocking seams on the edge of their panels, which make for easy installation.

Tile and slate are more difficult to install than regular shingles and they're subject to breakage if walked on after installation, or if some errant space junk hits your roof.

Most tile roofing pieces are designed to interlock with one another so that their own weight holds them in place. Your roofing sub and laborers first fasten small strips of wood called *battens* to the roof to provide spacing and support. They then set the tiles in place, starting at the bottom of the roof and working up. As the roofers work toward the top of the roof, they layer each level so that it overlaps the lower one. Artful cutting is necessary at seams and around chimneys and vents. At the top, special "peak" tiles are installed, which hang over both sides of the top of the roof. This layering forces water to stay above the tile surface and run over the roof rather than under it.

Gutters and drainage

Whatever your roof size or style, creating proper drainage is absolutely essential. If water backs up on your roof, it will soon find its way under your roofing material — and into your home. No matter what type of drainage system is installed, walk the property with your contractor to make sure the system is installed in a way that looks good on the house and works efficiently to remove water. The latter can be tested with a hose run on the roof to watch water drainage patterns.

To accomplish the goal of getting water off your roof quickly and efficiently, your roofers will flash your roof's valleys and angle your roof to channel water down the surface and toward the edge. If water is allowed to flow down to the ground directly from your roof, however, it will likely settle around the foundation, which can create settling problems in the foundation or moisture problems with the wood at your home's base. You don't want either of these outcomes to occur, and that's where guttering comes into play.

Keeping out of the gutters

Gutters can cause big problems if they get clogged with leaves or dirt. Water can pool or spill over, causing the very same problems they were installed to prevent. But, have no fear! New improvements are available today specifically designed to keep you out of the gutters! Some companies have designed gutters with special covers that allow water in, while keeping leaves and other foreign materials out.

Although these new-fangled gutters can be expensive, the cost (and the very real danger to you, if you decide to climb a ladder and do it yourself) of cleaning your gutters every year can add up — especially if your house is multi-story and requires extended time on ladders. Do some research on the cost of maintenance and figure out how long it will take you to make up the upfront cost in savings.

Gutters are channels — most often made of aluminum, plastic, or fiberglass — that collect the water that runs down your roof's surface and channel it into drainpipes at the corners of the house. Gutter subcontractors (who, by the way, generally charge by the foot) install the gutters with long, galvanized steel nails placed roughly every five feet. When installed correctly, gutters should be sloped down toward the drainpipe at a rate of 1 inch for every 12 inches of lineal feet. Your contractor can help you figure out the number of downspouts necessary and the methods of drainage at the foundation. Drainpipes connect to underground tubing that leads the water to your storm drain, or sometimes just to a concrete or plastic splash block that leads the water away from the house.

Installing the driveway — Finish options

Depending on your project's location and your contractor's preference, your driveway and sidewalks will generally be installed in the latter stages of your project. In rural areas, your driveway may simply be a gravel road; it may need to be created first so your subs can access the building site. In more urban areas, your choices can range from simple asphalt, which is a blend of tar and gravel, to luxurious stamped concrete — concrete that may be colored and that has patterns stamped into it while it's still wet.

Which material you choose for your driveway may depend, in part, on your neighbors. If you build in a rural area where houses are few and far between with no homeowner association restrictions, you may choose a less-expensive material for your driveway, like loose gravel. But if you build in a development, your community's codes may require you to choose the more finished look of asphalt, concrete, or even some type of decorative paver. Other choices of materials may depend upon cost and climate. Address the choice of driveway materials in the design process discussed in Chapter 5.

If you and your architect have designed a fancy driveway, keep this advice in mind: You may want to request that your contractor make the driveway the last thing installed before project completion so that construction traffic (those bulldozers, cranes, and heavy trucks can leave quite an impression!) doesn't destroy any of its beauty.

Concrete has become the most popular driveway material due to its durability and beauty. When installing concrete driveways and walkways, first the concrete sub builds wooden *forms* (wood boards that keep concrete within a certain area) in the shape of the driveway. The sub next sets in place metal *rebar* (reinforcing steel bar) on small blocks. The installed rebar looks like a net and it reinforces the concrete after it is poured. The sub then installs *expansion joints* (gaps between the slabs) to allow concrete slabs to move without cracking due to heat or cold (or earthquakes, in California). Before the concrete pour, the sub places and compacts a base of crushed limestone or recycled concrete to give the concrete a longer life. Your concrete sub takes into account the current temperature and humidity — vital elements in deciding exactly what concrete mixture to use — to assure the pouring of a clean slab. The driveway will be finished with a slight slope away from the house to allow for drainage. Cosmetic stamping, if desired, occurs soon after the concrete is poured.

Exterior lighting

After your home's exterior is finished, your electrician installs the outdoor lighting that is attached to the house. You and your architect selected lighting fixtures when you created your detailed materials list (see Chapter 5). No magic to finishing this part — the electrician simply turns off the juice and fastens those puppies in place. Because this task is fairly easy work, you do-it-yourselfers may want to participate in some of this installation process if your electrician is willing to let you help.

Make sure you don't have any standing water around when you turn on the power or you could get shocked! If you're installing halogen lighting, be sure to use gloves or plastic when installing the bulb. You don't want to touch it; otherwise the natural oil from your skin will significantly reduce the life of the bulb.

Spot-check — Exterior finishing

Now that your home's exterior is finished, your new neighbors can admire your home's good looks. To keep them admiring your home for years to come, use this spot-check to insure that those good looks won't fade fast. You need a ladder, a tape measure, and a hose connected to water. Be careful up on that roof!

❑ Siding is secure with no gaps or bare spots.

❑ Stucco is even and smooth. Stone or brick work is secure and in an even pattern.

❑ Deck is completely finished with no surface bubbles.

❑ Roof felt has full coverage and a vertical overlap of 6 inches. The roof extends over the eave by a minimum of 3 inches.

❑ Roof nails are pounded flush with no visible nails when standing on the roof.

❑ The entire roof doesn't have any loose or broken tiles or shingles. Patterns are straight and uniform.

❑ Gutters are secure, and drainpipes are attached to the wall every 6 lineal feet.

❑ Water sprayed from the top of the roof through a hose naturally goes into the gutters and out the drain system.

❑ Driveway is even and smooth with no noticeable cracks.

❑ Water runs off the driveway when sprayed with a hose.

❑ Exterior lighting is secure with no loose wires.

Discuss any issues and possible solutions with your contractor and subs.

Moving Inside — Completing Fireplaces and Walls

As soon as the outside of your new home starts shaping up, you can turn your attention to the inside, which means installing fireplaces, closing up and completing your walls, and prepping everything for painting.

Fireplaces and hearths

Most fireplaces today are prefabricated boxes that are readily installed in a space framed for this purpose by your carpenter. With insulated chimney shafts, there is little need for brick or stone other than for outside decorative or architectural considerations. A fireplace is best considered early in the building process because

✔ You need to order materials well ahead of installation.

✔ Your community may have restrictions on the type of fireplace, such as wood-burning or gas.

✔ If your house has gas and you decide to run a gas line to it (to fuel gas logs or to use as a fire starter), you'll have to run the required plumbing long before you get to this point in your build. We talk about gas lines in Chapter 13.

A custom masonry fireplace requires a proper *draft* (airflow from inside your home, up through the flue, and out of the top of the chimney) to function. A poor-breathing chimney can cause smoke to enter your house (causing your smoke detectors no end of trouble); too much breathing, on the other hand, makes keeping a fire lit difficult. Prefab fireplaces make life easy because they rarely have these problems.

Other considerations in a custom fireplace may be building a raised hearth to make access easier and an ash dump if the fireplace is on the ground floor. Some masons build in a fresh air vent allowing the fireplace to suck air from the outside of your home without sacrificing the warm air inside.

The last piece to be added to your fireplace is the mantel, that (usually) wooden shelf installed over your fireplace opening. A mantel can be custom built or bought prefabricated — many people acquire antique mantels from older demolished houses that can look quite nice. Make sure the mantel is firmly secured with fasteners before hanging your Christmas stockings.

Drywall and wall textures

Drywall is a kind of wall covering made of plaster or gypsum sandwiched between two layers of heavy gauge paper. Homebuilders commonly use it today to cover interior walls because it is uniform and easy to install. A few artists can still lay the old lath and plaster method of wall finishing (using wood strips and real plaster), but that method is generally too expensive compared to drywall, and doesn't provide any great benefit for the additional time and cost.

Drywall comes in two common thicknesses: ⅜ inch and ⅝ inch. The thicker board — ⅝ inch — is more commonly used to prevent *warpage,* or rippling of the finished surface. Warpage can occur in moist climates if you have studs spaced far apart, say 24 inches. You definitely want to avoid warpage. (Moisture is bad for drywall period, so make sure you store the drywall in a dry place on the site until it's installed — not out in your backyard in the pouring rain.)

Drywall is generally available in sheets of either 4 x 8 feet, 4 x 10 feet, or 4 x 12 feet. Drywall sheets have tapered edges on the long side and full edges on the short side. This way, when two tapered sides are put together, it creates a channel for drywall tape and mud to be applied, which are used to cover up and hide the joints between adjoining pieces of drywall. The result is a smooth, flat wall.

Working with a drywall sub

Your drywall sub is one the most crucial players in making your house look good. Discuss with your contractor about getting the best drywall sub you can afford, and you'll thank yourself a thousand times over when your custom home is complete. You can tell you have a good sub if she cuts pieces in place ensuring a tight smooth fit, if finishing coats are thin and feathered, and if the first two coats are so smooth that sanding is unnecessary.

Whether you want to hear it or not, a good drywall sub will likely tell you whether or not your framer did a good job. Knots, warped lumber, and out-of-square walls make her job more difficult, and your drywaller won't be shy letting you know about it. If the problems are significant, you may need the framer to fix them before the drywall sub can start.

You may have some additional choices to make at this stage that can cost additional funds. Discuss the following options with your contractor and drywall sub before work on these tasks begins:

- **Fancy textures:** Texture can add depth to a room, and the sub will have a selection available.

- **Unstippled ceilings:** Standard ceilings have a standard texture, but creating a smooth ceiling takes work and skill, which means additional cost.

- **Water-resistant drywall:** Over time, this drywall will hold up better in a humid climate or in the dreaded event of water damage from a leak.

Installing drywall

When your drywall sub starts to work, she'll generally install the drywall horizontally, with the full sides joining at the studs. Doing so reduces the amount of joints to be taped and places the strongest area of the panel across the studs. At the outside corners she'll install a metal strip with a beaded edge where the panels meet.

Your drywall sub covers ceilings in the same way as walls, except that if you desire a rougher *(stippled)* texture, she won't have to repeat the application of joint compound and sanding steps. Application needs to be done in moderate temperatures to ensure a quality job.

Spot-check — Fireplaces and walls

With walls and a fireplace, your home is starting to feel cozy. Check this list before moving on to be sure it's built to last. You need a flashlight, matches, newspaper, a Duraflame log, and a bucket of water.

❑ All bricks or stones on the fireplace have no cracks and are firm and secure, as is the mantel.

❑ The flue vents properly and will sustain a fire. Light a small Duraflame log and have a bucket of water handy just in case there is a problem.

❑ Rough cuts on drywall are cut close around window and door openings as well as plugs, switches, and ventilators so trim and plates cover properly.

❑ The drywall doesn't have any bumps or warpage. Look down the length of the drywall with a flashlight to verify. Circle flaws with a pencil for repair.

❑ No corner bead is exposed.

Discuss any issues with your contractor and determine when and how to reconcile them.

The Finish Carpenters — Doors, Windows, Molding, Cabinets, and Countertops

No, the finish carpenters aren't from Finland, but they are responsible for the finest details in your home, which determine whether your home is ordinary or extraordinary because you (and your friends, business associates, and relatives) can see their work everywhere.

Your contractor will likely have long-established carpentry subs, but if you have a choice in selecting carpenters, choosing this subcontractor should entail a very deliberate process of examining your candidates' skills and workmanship before you even think about making a hire. Make sure you see as many previous samples of this person's work as you can before you and your contractor decide to give him the job.

When you and your contractor select the finishing carpenters, remembering what types of jobs each sub will perform on your house can make the decision easier. The following is a list of jobs typically completed by your finish carpenters:

✔ Hang interior and exterior doors

✔ Set door sills

✔ Set windows and window sills

✔ Install closet and pantry shelving

> ✔ Install stairway trim and decorative railing
>
> ✔ Install base and crown molding
>
> ✔ Install decorative molding such as chair and picture rails and wainscoting

When working with your contractor to select your finish carpenters, keep in mind that they may figure their bids in different ways. Some calculate their job by the number of openings cut into the wall (for example, the number of windows and doors) or the number of cuts they have to make. Others charge on a time basis by the hour or by the day. You can save money by having your contractor negotiate for the whole job and then work to make as few changes as possible.

After your contractor has selected the finish carpenters, make sure you and your contractor have planned your finish completely. Ideally, all your materials need to come from the same suppliers — this way, everything matches. If you run out of trim during the course of the build process and have to order from a different supplier, you may find slight variations in the color, grain, weight, or other aspects of the materials.

After you've ordered your materials and they've arrived, remember that you'll be storing a lot of trim material at the site, including moldings, rails, doors, and more. Make sure the material is stacked in a dry, out-of-the-way place where it won't be damaged from foot traffic or by the outside elements during the construction process. Scratches, nicks, and broken glass can cost you extra money due to repair or increased waste and material cost.

Doors and windows

Here's a bit of advice from someone who has hung a *lot* of doors: Doors must always be hung level or *plumb,* and the door trim should also be level and square. If your sub fails to take this essential step, then we can almost guarantee that you'll have problems closing your doors at some time in the future — hardware may not match up and the door simply won't work properly. Make sure your sub takes special care when creating the correct clearance at the bottom. If doors are installed incorrectly, they can stick — taking a lot of effort to fix.

Your exterior doors will be solid and in many cases decorative. Weather stripping is an essential add-on for exterior doors because it saves energy. Discuss with your contractor who will be providing the weather stripping because it may not be the finishing sub.

You have choices with interior doors that depend upon price and the look you desire. At the bottom of the price barrel are hollow-core masonite doors. These doors have an outer frame with a sheet of masonite on either side.

They're light, inexpensive, and easy to hang but must be painted and don't provide any real aesthetic benefit.

Alternatively, you can choose from many prefabricated hollow-core doors today with imprinted three dimensional patterns and simulated wood grain. These doors still require painting, but can add a nice decorative flair and are simple to install. You can also purchase some hollow-core doors with wood veneer that can be stained.

Custom doors will set you back in price. These doors can be beautiful, but solid-core doors are heavy and need expert hanging. Furthermore, they need to be ordered far in advance to make sure they're at the house in time for the finishing sub. Your door decisions are best made during the architecture phase of your custom home project, as discussed in Chapter 5. Doing so allows your contractor and finishing sub to make appropriate arrangements in advance.

Prehung door and window frames can make life much easier and can be easily fit into place. Most of today's windows are made from metal, wood, or vinyl and are completely self-contained units. If your windows weren't installed immediately after framing, discussed in Chapter 13, your carpenter now secures the door or window unit in place and then levels it with small flat wedges of wood called *shims*. When the unit is level and secure, he installs flat or decorative trim called *casing* around the window or door to hide all the framing and make it look uniform.

Baseboards and moldings

Baseboards (molding that runs along the corner of the wall and floor) are relatively simple for your finishing carpenter to install. He simply seats the measured and cut boards against the floor and secures them onto the wall with a nail gun. (Gone are the tedious days of carpenters pounding in finish nails and then painstakingly sinking each nail below the surface with a small metal punch.) The corners of the baseboard will likely be *mitered,* meaning they're cut at angles so they fit smoothly and cleanly into place.

You have other finish molding options to consider, such as:

- **Chair** or **picture rails** are decorative molding strips that run somewhere along the middle of the wall.

- **Crown molding** runs along the corner of the ceiling and the wall or sometimes just below the ceiling. Crown molding is more difficult to apply because most decorative crown sits at an angle to the wall, which can make negotiating corners and angles difficult.

These options require a little more expertise to install and will require additional costs.

Keep in mind that, because these moldings are in direct sight, critical neighbors and mothers-in-law will probably scrutinize them more. The almost-lost art of coping is the solution. Although most finishing subs can miter joints in their sleep, it takes real talent to hand carve molding with a coping saw to fit perfectly. If your trim is finished wood rather than painted, finding a finishing sub capable of this skill is critical.

Another challenge with moldings (which are generally flat) is dealing with curved walls. A good carpenter can work angles almost into curves — an absolute necessity for wood grain trim. Custom warped pieces can be made but are very expensive. If the trim is being painted, new flexible moldings are available today that can bend around complex curves. Discuss these choices with your architect in the planning stages so there will be fewer surprises when you're so close to moving in. (See Chapter 5 for more design topics.)

After the trim is in place, your finishing carpenter fills in any nail holes with putty. After the putty dries, the carpenter sands and then stains or paints the nail holes.

Your finishing carpenter and laborers also complete railings and *spindles* (the individual spokes) on your staircase. They work painstakingly to secure each piece before applying the *banister* (the top of the railing), and sanding and finishing the railing. Your carpenter also finishes the steps and *risers* (the facing pieces of your stairs).

Cabinetry and countertops

Cabinets and counters are installed in the most expensive rooms of the house — the kitchen and bathrooms.

The bad news? You'll sink a lot of money into these functional decorative elements. The good news? You can derive great value by investing in good quality and pleasing design. Prefabricated cabinets are common today, and they can be made in excellent quality and accommodate custom sizes and needs. Buying prefab cabinets can be significantly cheaper than having a carpenter build your cabinets from scratch, but depending on the look you want and taking into account the intricacy of the cabinet areas, custom may be necessary or desirable. Discuss your options with your architect when making choices about wall shapes and areas.

You'll have many choices to make before ordering, including shapes, styles, and colors of your new cabinets. You can also choose from a variety of options including inserts, pull-out drawers, lazy susans, dish and spice racks, trays, and more. Are you having fun yet? If using prefabricated cabinets, most of these specifications are chosen early in the process with your architect or contractor.

When ordering prefabricated cabinets, you may get a significant discount from your sub if you order all the house cabinets from one manufacturer. If you have chosen relatively simple stock designs, some bathroom cabinets may come with preinstalled counters affording you savings in time and money. Discuss these options with your contractor and sub.

Just as you have an almost unlimited number of options available for cabinets, you also have an almost unlimited number of options available for countertops. For years, the kitchen countertop of choice was Formica. Today, however, those old Formica tops are looking pretty dated, and many custom home owners are instead opting for the long-lasting beauty of granite, marble, or tile. You can also choose from several synthetic choices such as Corian and Silestone. Your budget and taste generally determine which one you select.

In your regular meeting with your contractor, make sure that the appropriate subs have been scheduled correctly so that, for example, your tile sub starts work *after* the cabinets have been installed instead of *before*. (Check out Chapter 11, which explains the importance of having regular meetings with your contractor and what you need to discuss.)

Reconfirm the measurements and layout of your kitchen and bathrooms with your cabinet sub. Make sure he has the dimensions of the appliances as well as the placement of windows and doors. This information insures that everything fits in all the right places. Walk through all the cabinet areas with your contractor and sub immediately after framing to make sure everyone is in agreement on the plan. You can mark on the floor or studs with chalk. Your contractor makes sure the appliances are installed correctly so that your carpenter can finish the areas around them.

The following list provides some typical dimensions to remember as reference. These are standard dimensions. Any variations on these measurements require custom ordering and installation (which costs you a *lot* more money). Try to avoid them if possible.

- 36 inches: The countertop height from the floor
- 32 inches: The optimum height for built-in desks and writing tables
- 24 inches: The knee space required underneath desks
- 24 inches: The standard countertop depth
- 3 to 4 inches: The standard back-splash height behind sinks

If you schedule your walk-through before the drywall is installed, you need to remember to account for the thickness of the drywall as well as the width of molding to be added when measuring for cabinetry. Check these numbers with your contractor carefully.

Some contractors feel cabinets should be installed and then stained or painted for the best look. Others believe before any of the cabinets are actually set in place, they need to be stained, painted, or otherwise finished. Doing so makes sure that otherwise hard-to-reach areas are fully covered and will keep stain or paint from getting onto unwanted surfaces. Discuss the options with your contractor.

Upper cabinets for the kitchen are installed first. Why? Because putting them up without the base cabinets in the way is easier. After the base cabinets have been installed, the carpenter can cut countertops and fasten them securely in their proper positions. Next, the carpenter cuts appliance holes (but make sure your finishing carpenter cuts the holes when the appliances are actually on-site, so you can double-check the dimensions). After the carpenter adds the back splashes, he caulks all seams and joints, adds drawers, and attaches pull knobs. He also repairs any scratches or marks. Sometime during this phase of your project, the finishing carpenter also installs bathroom vanities. Check with your contractor to schedule the plumber so he can connect the sinks, drainpipes, and faucets after installation.

Spot-check — The finish carpenters

Are your carpenters doing the right stuff? Or are they just a couple of singers named Richard and Karen? Check this list before moving on. You need a tape measure, level, and flashlight.

- ❏ All doors open and close freely. Locks work smoothly. A ¼-inch clearance exists above carpet or threshold height.
- ❏ Windows are level and open freely with no squeaks or binding. Screens are installed.
- ❏ Weather stripping for both doors and windows is securely in place.
- ❏ All marks and nails are covered and finished.
- ❏ Staircase railings and spindles are secure and smooth.
- ❏ All cabinet doors open and close without squeaking. When opened halfway, they stay put.
- ❏ Cabinet finishes are even, not blotchy.
- ❏ Countertops are secure and level. Place a marble on the surface to see if it rolls (if it does, there is a problem).
- ❏ All drawers work properly and can be easily removed and replaced.
- ❏ All joints are caulked and built-in appliances properly fitted.

Discuss any issues or discrepancies with your contractor and determine possible solutions.

All the Pretty Stuff

This section covers the D in detail — all the little items that people often take for granted, but that stick out like a sore thumb if they aren't done right. If you have updated or remodeled a home in the past, then you have probably worked with some of these processes before. Many of the details in this section are standard home-improvement techniques and, aside from this overview, you can find more information and specifics in *Home Improvement For Dummies* by Gene Hamilton and Katie Hamilton (Wiley).

Painting

Having the interior painting finished before the flooring goes in is a good idea. Believe us, nothing is worse than having a big, sticky paint or polyurethane accident on your brand-new hardwood floors or carpet.

Many people like to consider doing the painting themselves. Unfortunately, not everyone is truly a Rembrandt, and it takes a lot of time to paint a house. Consider the following:

- ✔ Do you have the skill?
- ✔ Do you have the time?
- ✔ Do you have the physical ability?

If you can comfortably say yes to all three of these questions, then you can certainly have some fun doing the painting while saving a little money in the process. If you're busy with work, or if the last time you painted was with your fingers or little Susie's pigtails, then this work is probably best left to professionals.

If you decide to hire a painting sub, discuss with your contractor the need for hiring the best painter with great references *and* insurance. Try to see samples of his or her work on other home projects if you can. Remember: Your walls are the finished surface that *everyone* will see. Don't be cheap about the artist or the materials.

Your new house has additional needs because it has never been painted before. For example, new drywall needs a coat of primer and two coats of paint to seal and cover the first time — even with guaranteed one-coat paint. Your painter needs to be a true professional to be able to anticipate and deal with these special needs.

Different surfaces have different finish requirements. Here are standard surface and finish choices:

- Walls in bedrooms and living areas — Flat finish

- Doors, trim, and woodwork — Gloss or semigloss because they repel fingerprints

- Kitchens and bathrooms — Gloss and semigloss because they're easier to clean and repel moisture

Another common question for the happy homeowner is what type of paint to use. You basically have two choices in materials. Both can be obtained with a variety of finishes. Discuss the best choice with your contractor and your painter.

- **Oil:** Oil used to be the paint of choice because of its durability. It bonds well to wood and metals and works well on oily or dirty surfaces.

  Oil can be a problem in damp climates, however, because the moisture repels the oil base during application. Other disadvantages of oil are that brushes and spills can only be cleaned with thinner, and it smells bad until it dries. Oil is generally considered the best choice for the exterior, but not for the interior.

- **Water-based paints, such as acrylic latex:** Acrylic latexes have become extremely popular due to their ease of use. They are durable and safe, and cleaning up brushes and spills during the application process is easier. And because these water-based paints dry much faster than oil-based paints and produce no fumes, they can be the best choice for interior walls.

  When considering water-based paint for exterior use, be aware that it is a porous paint, meaning that moisture can penetrate it when applied poorly over wood without an oil-based primer. If you have a stucco exterior, acrylic latex may be a good option because the stucco acts as its own sealer.

 If you're looking to have specifically designed colors throughout the house, or are considering murals or faux finishes, you may need the help of a designer before engaging your painting sub. If you want to make the decorating decisions when everything else is done, you can have your painting sub finish everything in white for now; however, you'll find it much cheaper and easier to have all your painting done (including the custom work) before you finish flooring and move in furniture.

Hardware and fixtures

People take for granted all the hardware and fixtures that must be installed after everything else is put together. In Chapter 5 we spell out all the hardware and fixtures to choose for your home; now, during the finishing process, is the time for your subs to put them in their permanent place. Here's a to-do list for your hardware and fixtures. Some of these you may want to do yourself.

- ✔ Install hinges on all doors.

- ✔ Install doorknobs on all interior doors.

- ✔ Install matching deadbolt and lockset on front and back doors.

- ✔ Install light switch and electrical cover plates.

- ✔ Install all lighting fixtures and ceiling fans.

- ✔ Secure and attach thermostats and alarm controls.

- ✔ Install all exhaust fans.

- ✔ Install toilets and toilet roll holders.

- ✔ Install towel racks and soap dishes.

- ✔ Install medicine cabinets and mirrors.

- ✔ Install shower rods and doors.

- ✔ Install faucets and connect stoppers.

- ✔ Install window locks.

- ✔ Install doorbell.

- ✔ Install door knocker.

- ✔ Attach all cabinet knobs and appliance knobs.

If you're installing lighting fixtures yourself, be sure that the circuit breakers have been shut off first. This installation is simple because most light fixtures come with clearly marked wires and printed instructions. Be sure to use only recommended wattages because creating an over-wattage lighting system can generate excess heat and even cause a fire. At the very least, it will cost you more money on your monthly energy bills.

Flooring materials

You made choices about your flooring materials in Chapter 5 and, by now, you have ordered your materials and they're ready to install. For you do-it-yourselfers, flooring installation is a task where you may want to have some fun. Decide with your contractor and flooring sub how you might participate in the process.

Install the flooring after the painting has been finished. Before any flooring is laid, however, your contractor makes sure the job site has been cleaned with set traffic pathways to avoid dirtying the new floors. To prepare for flooring installation, the subfloors should be cleaned of any rough spots or plaster. Depending on the type of flooring and location of the subfloor, your sub may install a waterproof vapor barrier. Here are some insights on installing common flooring materials.

A bright idea

Most people think Thomas Edison was the first inventor of the electric light. Actually, he was the 23rd inventor of the electric light. Sir Humphrey Davy actually invented the arc lamp in 1811 and Sir Joseph Wilson Swan had already constructed a glowing bulb in the mid 1800s. But that bulb had no vacuum and couldn't be sustained. Edison actually is charged with producing the first vacuum bulb that could sustain light and be mass produced. But this was only the spark for Mr. Edison's real contributions.

Because gas lamps were the order of the day, houses weren't rigged for any electrical appliances. Therefore Edison could only sell light bulbs by creating the sockets, wiring, plugs, and whole electrical infrastructure necessary to bring power into the home. (This is why standard two pronged plugs are referred to as Edison plugs.) The need to sell his many inventions led to the creation of his power companies and electrical supplies company. This firm was originally known as Edison General Electric and today is known as General Electric. Of the original 12 members of the Dow Jones, only GE has survived to this day. You can even thank Mr. Edison for an invention that makes you cringe every month, the electric bill. For more interesting info on Thomas Edison, check out the Edison Preservation Foundation at www.edisonpf.org.

Wood and laminate

Many people love the natural beauty of wood floors, and the installation process is fortunately relatively simple. Your flooring sub will move briskly through the process.

Your flooring sub first measures and cuts straight planks, and then either glues or nails them into place — or both. *Tongue-and-groove flooring* gets its name from the protruding wood on one side of the board that fits into the slot on the other side of the board. If you're using this type of flooring, your flooring sub insets piece by piece and allows ¼-inch space alongside walls to allow for moisture expansion. If you're using *parquet flooring* (flooring with an inlaid design), the flooring sub usually lays it down in 6-inch squares, which allows for the same breathing room.

To avoid major headaches, subs will try not to lay hardwood floors in humid weather, which can create gaps in the boards when the weather dries out and the flooring contracts. Most contractors have the flooring delivered to the house a few weeks early to give it time to dry out. You can then store the flooring in the actual rooms in which it will be laid to let the moisture content stabilize. After the flooring has been laid, sanding may be required before its final finish.

Laminate floors like Pergo and WilsonArt have gained popularity due to their ease of installation, their durability, and their versatility of style. Some are completely waterproof and have lifetime guarantees, and some are amazingly accurate in their representations of wood and stone materials. Many laminate floor products are available today for do-it-yourselfers. The material the manufacturer is imitating (wood or slate, for example) determines the size and shape of the panels. Most stone imitations are square, while the wood may imitate planking. Panels are generally designed to interlock and can be applied with glue directly to the slab or subfloor.

Vinyl

Like laminates, vinyl floors have great versatility of style and durability. Vinyl floors can come in rolls or squares and are very easy to install. If you're installing the floor yourself, make sure you have planned out your patterns so they make sense with the angles of walls and placement of doorways. Also, be sure to follow specific directions supplied with the flooring so you don't void the manufacturer's warranties. Vinyl floors are generally installed using a special adhesive made specifically for this purpose, although some products may be nailed in place or may even be self-adhesive.

Tile

Tile flooring can be beautiful if done right, but it takes talent — and a lot of patience. You seriously need to consider your craftsmanship skills before embarking on a tile mission yourself. If you have any reservations, then by all means use a pro.

Tile first needs to be properly cut and spaced when being applied. After you approve the pattern design, the sub applies a thick sticky glue, sometimes to a special, inflexible concrete-impregnated surface called Wonderboard (used to firm up a flexible wood floor). The sub then sets the tiles in place with small plastic spacers to fix their position. After the tiles are secure, the sub fills the space in between the tile with a gritty colored material called *grout*.

Darker colors of grout show dirt more than lighter colors, so be sure to discuss the color options with your tile sub. Your tile sub may recommend a sealer to protect the grout and make it easier to clean.

Carpet

Carpeting may also not be the best job for doing yourself — chances are that a professional will do a much better job. Something to remember: Buy the best padding you can afford. It keeps your carpet feeling good and helps it last longer.

Carpet is sold in 12-foot-wide rolls. To save money when carpeting small areas, ask about *remnants* (pieces left over from large carpeting jobs, sold at a discount). Carpet is stretched upon installation by attaching it at the walls to a wood strip filled with tacks called a *stretcher strip*. After a few months, it may need to be stretched again. Make sure your carpet company arranges for this without an additional charge.

Appliances

Some of your appliances will be built-in appliances as opposed to free-standing and should be installed at the same time as your cabinets. The free-standing appliances can wait until all the painting is finished and flooring is installed.

Spend time with your contractor — at your regular meetings we discuss in Chapter 11 — to determine the proper time for delivery and installation of the appliances. Having them delivered too early risks damage and can create obstacles to work around. Having them delivered too late can delay your move-in schedule, or worse, the completion of your construction loan. Make sure upon delivery that the measurements are as specified in your building plans.

Spot-check — Painting, fixtures, flooring, and appliances

Time to check all the detail work. Don't forget: This phase makes or breaks your project. Take the time to look over the detail work in *detail!* You need a tape measure and flashlight.

- ❑ All ceilings and walls are uniform in color with no brushstrokes or dried paint drips.

- ❑ Trim is painted with gloss or semigloss and is clean of bleed-over on walls or window glass.

- ❑ All hardware and fixtures are in place and secure. Knobs are tight and uniform.

- ❑ No carpet seams or gaps are visible. Vinyl flooring is bubble-free.

- ❑ Wood flooring is stained evenly with no visible scratches or nicks.

- ❑ Tiles are secure and evenly spaced with grout running parallel and perpendicular to walls. Tiles aren't chipped and colors are consistent.

- ❑ All necessary door thresholds are installed and secure. Doors are at proper height with no more than ¼-inch gap under interior doors.

- ❑ All appliances are securely fastened and plugged in. Gas and water connections are checked for leaks.

Discuss any discrepancies with your contractor and determine a solution.

Part IV
All the After Stuff

In this part . . .

Breathe a sigh of relief. Your new home is done and you can move in, though you're not quite done with all the decisions yet. In this part, we explain how to wrap up all the bank and government paperwork. We also inform you of moving options as well as refinancing and other financial strategies associated with your new house. Finally we give you guidance through the landscaping and hardscaping issues that you need to complete to turn your house into a home.

Chapter 15

Home Sweet Nest Egg: Moving In and Managing Your New Investment

Congratulations! Where once was bare land now sits your beautiful custom home! All the construction workers have done their job, and you've seen all your dreams come to life. Your home may have cost a bit more than you expected. It may have taken a little longer than you thought, as well. But chances are it was worth every dime and minute spent — at least we hope so! However, the project isn't quite finished. You now need to wrap up the loose ends.

In this chapter we walk through the last details of dealing with permits and getting ready for your move-in. We examine the issue of repairs and contractor warranties and discuss the pros and cons of staying in your new house.

Finishing Up the Custom Home Project

The process of finishing your new home is a combination of action and paperwork. Of course, you need the house to be ready for you and your loved ones to inhabit, but it also needs to be ready for the bank as well as the city or county. Unless you chose to be your own contractor as discussed in Chapter 1, you'll heavily rely on your contractor to help you through this final process.

Getting your certificate of occupancy

Obtaining your certificate of occupancy (C of O) is the first step to finishing your house. City and/or county building inspectors have been visiting your property regularly to inspect and sign off on the foundation, electrical, plumbing, and other key components. (If you want to know more specifics about the inspection process, check out Chapters 6 and 11.) When the house is nearly finished, the building inspector will come out for the final inspection. In most areas, the inspector isn't concerned with minor deficiencies in workmanship, quality, or cosmetic issues such as paint, flooring, or window covering. He is focused on making sure that the house meets all the structural and safety guidelines required by the local building code.

The contractor generally calls for this final inspection a few days in advance of being ready. Take the opportunity to walk the house with the contractor the day before the building inspector arrives. Have the contractor explain and show you all the items the building inspector will be looking at. This information can give you the chance to clearly understand the process and check up one last time on the contractor's work. If you see any defects, this is a good chance to point them out so the contractor can have the subs fix them.

After the inspector reviews the house and is satisfied that it meets all local building codes and requirements, he files the approval with the county or city. The building inspector's signature of approval allows the building department to issue the C of O. This document allows you to move in and will be critical for other parties, such as your bank and the tax assessor. It means that your house officially and legally exists.

If the inspector decides the house isn't finished, you have two options:

- ✔ You can wait until the contractor performs all necessary tasks to meet the final inspection.

- ✔ You can have the inspector issue a *temporary C of O* allowing you to live in the house while the necessary tasks are completed. (A temporary C of O can be beneficial if you've already sold your original home and scheduled moving trucks and utilities.) This option may not be available in every area of the country; check with your local building department.

We discuss the moving process in the "Selling the Old Home and Moving" section, later in this chapter. In either case, the building inspector has to come back to the house to inspect all the work.

Obtaining the mechanic's lien releases

After the C of O is in hand, you need to wrap up all the house's financial details. As we discuss in Chapter 11, all workmen have the right to attach *liens* — legally enforceable financial claims — to the property if they aren't paid for their services. Many of your suppliers and subs will have issued *preliminary filings,* which put everyone on notice that work has been done or materials have been delivered to the property. *Preliminary filings* — which allow the contractor and subs to file their mechanic's liens instantly if you don't pay them — have been happening during the entire construction period. In order to eliminate the risk of these preliminary filings turning to mechanic's liens, you need to obtain a *lien release* from each and every sub and supplier as well as your contractor. They'll only issue the release if everything has been paid for accordingly.

The best way to handle lien releases is to collect them as you move through the process. If the contractor has been collecting them, make sure she provides you copies along the way. Keep careful files so you have the lien releases immediately available when needed.

Mechanic's liens can be problematic for your loan and your finances. Our experience: Don't dispute contractors or subs by withholding their funds. Pay them if possible and sue them after the fact. Doing so allows you to clear the title on the property and resolve the disputes independently.

Rolling the construction loan — Choosing a final loan amount and program

Time to finish up the financial issues. Most likely, after reading Chapter 8, you chose a single-close, all-in-one construction loan and are ready to roll it to a permanent mortgage. This type of loan requires a certain amount of paperwork to get the lender moving on the conversion process. Here are the items your lender will need to start the rollover:

✓ **Final draw request:** Your lender probably set aside 10 percent of the funds in the loan until everything was finished. You can now request the money. For more information on how the draw system works, read Chapter 10.

✓ **Final lender inspection:** After you have requested the last of the money, your lender will want the inspector to come out and make sure the work has been done according to plan. Be aware that the building inspectors aren't looking at the work quality. See Chapter 11 for more details.

✓ **Copy of recorded C of O:** We discuss the certificate of occupancy in "Getting your certificate of occupancy," earlier in this chapter. The lender doesn't finish the loan without it.

✓ **Lien releases:** The lender's security is at risk until all contractors and subs are paid and have released the property. You need everyone to sign a release.

✔ **Verification from the title company:** The property is the only guarantee the lender has in case you default. Your lender wants the title company to reaffirm that the property is still yours and that no other liens or restrictions take precedence over the loan.

✔ **Evidence of insurance:** Now that you have a home, you need to obtain a homeowner's policy. Call your insurance agent and tell him you need a policy covering replacement value for the improvements.

As soon as you've collected all these items and given them to the lender, you may have the option of changing loan programs or changing your interest rates — of course depending on your lender and the loan program you selected. (Check out Chapter 8 for more details on these choices.)

If the final terms of the loan are insufficient for your needs or if you're rolling from a straight construction loan, you still need all these items to pay off the construction lender. Furthermore, you also need to find a lender for a refinance. We address the details of this process in Chapter 16.

Taking the Final Contractor Walk-Through

The C of O is in, and you're wrapping up your lender's requirements. Make sure the house is finished to your liking and the contractor is finished when you take the final walk-through.

A custom home is a huge project with many parts. Only in rare cases is everything done absolutely perfectly the first time. Ideally, you have been communicating with your contractor regularly and spot-checking along the way. We provide you with spot-check lists in Part III exactly for this purpose. Now take the time for the final walk-through.

Make sure you have plenty of time and a flashlight because you and your contractor are going to go through every detail in the house. Along the way you and the contractor write down every imperfection to be fixed. This list is called a *punch list*. Don't feel rushed during this important part of the building process; plan for several hours of sniffing and snooping. Look above, below, and behind everywhere you and your flashlight can reach. Have the contractor sign the punch list at the end of the inspection. (If you do notice issues, check out "The builder's long-term responsibilities and warranties," later in this chapter, about how to get them fixed.)

You'll hear many people advise you to withhold funds from the contractor if he isn't performing to the standards you require. Unless you have agreed in the contract to withhold some funds until the punch list is complete, Kevin suggests, based upon the many projects he has witnessed, to not use the money as a lever against the contractor. Most contractors run on slim margins and if they file a mechanic's lien, it can cause considerable hardship for your project especially if you have a construction loan. Kevin says the best way to handle the situation is to keep communication open and friendly, resorting to legal action only when absolutely necessary. If the contractor is truly in the wrong, the legal system will force the contractor to cure the problem, and you should be able to recoup your legal costs. Rarely has Kevin seen conflicts go to the point of lawsuits because contractors have their license and livelihoods on the line. We discuss preventative measures for these situations in Chapter 11.

The contractor needs to arrange for all the repairs on the punch list to be made in a reasonable time frame. He works to have the serious and messy repairs done before move-in, but you may have repairs being done for some time after you've moved into the house. The contractor may have to order materials or work with subs' schedules to get the work done.

Selling the Old Home and Moving

The most important issue with selling your old house is the timing. Gauging exactly when the new house will be finished and ready for you to move in is difficult. At the same time, market conditions fluctuate, which can affect finding the right time to sell. The ultimate goal is to have as little time as possible where you're carrying mortgage payments on two houses. You also don't want to be too quick on the draw, where you take an overly low price. Furthermore, if you sell the home too early, you could have no place to live or have to pay for storage and a double move.

Our advice is to carefully watch the market and discuss the options with your real estate agent. If you aren't using a real estate agent, you'll have to assess the market yourself by talking to real estate agents. Find out specifically how long it will take to sell and close your home. Build contingency plans for renting out the house if necessary. Chapter 7 describes how to use your cash resources to resolve any cash-flow issues in the beginning, leaving you the relaxed atmosphere to take advantage of the right sales opportunity. Meanwhile, stay in constant communication with your contractor as to when he thinks the house will be ready for move-in. In our experience, you must add two weeks to this date just to be safe. The worst case is that you'll have an extra two weeks to make the move. The alternative could get ugly.

The moving process itself can be long and complicated if you're doing it yourself, which is why so many people opt for movers. When moving to a custom home, you have the benefits of time and knowing what the house will look like. This advance notice beats finding a house and moving in 60 days, the typical time it takes to close a deal for purchasing an existing home. The Web has great resources for movers and moving tips. Try www.move.com or www.

uhaul.com. Aside from all the general moving tips available today on the Internet, the following list includes some of our suggestions for making a smoother move to a custom home:

- ✔ **Use the floor plan.** Take advantage of your plans. You can make cutouts to scale of all your existing furniture. Doing so can tell you where everything fits and allow you to weed out all the stuff that doesn't belong.

- ✔ **Do some advance decorating.** As you see the house come together, you should already be thinking about colors and decorations. After you have a sense of the rooms, you can assess the items in your current home that will stay or go.

- ✔ **Use pictures to see if items belong.** Chances are that you'll buy a lot of new stuff for this house. You can start early and get measurements as well as photos from the retailers or the Internet. Try and have deliveries set up for a time just after the move.

- ✔ **Rent a dumpster.** Most of Kevin's clients start almost from scratch when decorating their new home. We can think of nothing worse than paying to move something and then paying again to have it hauled away. Be very selective in what you keep and get rid of everything else.

- ✔ **Have plenty of garage sales.** We can't think of a better way to raise capital for all the new furniture you'll likely buy. And there's nothing like the feeling of a big wad of cash in your pocket at the end of the day.

Managing Maintenance and Repairs

When building your new home, you probably hoped that it would require a minimum of repairs and ongoing maintenance. Think again! For example, compare it to manufacturing a car. The most reliable cars are ones built with robotic factory precision. The bugs have been worked out after years of manufacturing thousands of copies of the same vehicle over and over.

Keeping with this analogy and comparing it to your house, your house isn't factory made. Your new home is like an expensive custom car. The components have been picked one by one, and they have been handcrafted together to give amazing performance and luxury in state-of-the-art surroundings. Ideally, everything stays carefully tuned, but chances are you'll encounter some minor glitches in the first couple of years or so.

All the equipment and appliances have warranties, and the contractor guarantees her work for a period of time that varies from state to state. Make sure the contractor gives you all the manuals and registration certificates for the appliances and fixtures. Catalog them and store copies in a place away from the house in case of flood or fire when you'll definitely need the information. You can catalog and easily access many registrations online.

The builder's long-term responsibilities and warranties

Contractor warranty policies vary from state to state. A typical ten-year warranty covers structural defects for ten years, as well as construction materials and defects for the first year and major mechanical systems for the second. After you get past ten years, any problems are going to be yours unless they boil down to significant engineering issues or outright fraud.

Discuss with your contractor the full details of his warranty, including who covers the work if he goes out of business. Have him walk you through suggested maintenance procedures for keeping the house in working order.

Dealing with construction defects

Okay, so not every contractor is perfect, and your house may have some accidental (or in rare cases, intentional) defects in workmanship. Sometimes you don't catch these defects in the final walk-through. In fact, they may not show up for years. A perfect example is a cracked foundation that doesn't become apparent until years after the home is completed, or drainage or roofing issues that don't materialize until the first big rain. These types of construction defects are a big nightmare for the homeowner, and resenting the contractor for his negligence is only natural.

Do yourself a favor. Take it slow. If you find yourself in a construction-defects situation, you don't want to escalate things unnecessarily. Remember that your objective is to solve the problem with the least hassle and expense. Even though ultimately you may win a war with the contractor, it could come at the expense of costly emotional and financial battles. Often the contractor's contract covers issues of this nature. Make sure you discuss these issues when you're hiring the contractor (see Chapter 11). Don't worry if the discussion got missed somehow; you can work out disputes amicably by following this path:

1. **Talk to the contractor calmly.**

   Have the contractor come out to the house and see the problem firsthand. Discuss how it could have happened. You may need to involve the architect and engineer if the problem is significantly structural in nature. Keep discussions calm and cool in determining how to remedy the situation. Most contractors want to find a way to solve the problem. They want to protect their licenses and their livelihoods. If you have picked a contractor with a good reputation, chances are he'll want to keep it that way. On the other hand, if you aren't getting anyone to step up for repairs or they're all playing the blame game, go to Step 2.

2. **Get an outside opinion.**

Getting an outside opinion isn't cheap, so keep track of your expenses. You want to hire an expert in the field in question. If the issue is cosmetic, such as the cabinets, bring in a finish carpenter. For structural issues, hire a structural engineer. Be prepared to spend a few hundred dollars to get an expert opinion for submission in court if necessary. Make sure that the expert is willing to testify if necessary. Get two independent opinions if possible. Doing so may cost you twice as much, but it'll give you a better case. Make sure you get two or three outside estimates for curing the problem as well.

3. **Suggest mediation.**

Contact all the pertinent parties. Arrange a meeting at the house where everyone can once again see the problem. Show the independent reports and estimates to your contractor, architect, and engineer, if involved. Take the time for negotiation. You may get lucky, and your contractor may agree to fix the problem at no cost to you. If so, make sure you get it in writing just like any other contract for work, complete with the time frame and a statement that the work won't be at any additional cost to you. Even if the contractor suggests some cost to you in the process of negotiation, consider this cost against the expense of an upfront attorney and protracted litigation. Make sure you point these issues out to the contractor and subs, if involved, as well. If they still don't budge, move to Step 4.

4. **Get an attorney.**

Okay, last resort time. Time to do battle. The big mistake in this step is picking the wrong attorney. Stay away from general practitioners who dabble around in various aspects of law. Your family attorney may be fine for small liability and estate stuff, but the laws on construction defects are very specific. Kevin once saw a well-meaning attorney negotiate his client into a series of bad compromises leading to hundreds of thousands of dollars in losses, all because he didn't know the field. Unfortunately, most attorneys don't know when to say no if they think someone has been wronged, and they end up making matters worse. Bottom line: Find a specialist in construction defects and be prepared to spend two to three years and $20,000 to $30,000 in legal costs before you can get satisfaction from the courts or a settlement. If the builder remains solvent, you should be able to recoup your legal costs as well.

Should You Stay or Should You Go?

You have gone through all this pain and hardship to build your new home and it's finally done. You have moved in, landscaped, and decorated. (We hope you referenced Wiley's *Home Decorating For Dummies,* 2nd Edition, by Katharine Kaye McMillan and Patricia Hart McMillan.) You can now take a deep breath and relax.

Who in their right mind would even think about moving now? You'd be surprised. For many people, the project and process part is more fun than the actual living in the house part. Many people also gain financial appreciation in the process and have the opportunity to walk away with a nice bit of money. Sometimes work-related issues can force a relocation. For whatever reason, you need to consider certain issues related to selling a custom home.

Being aware of neighborhood trends

Finding the right time to sell in a custom home neighborhood can be a little tricky. Remember that neighborhoods are designed to mature in 25 years. When you sell in a new neighborhood, you're selling a home to someone that wants to live in an undeveloped neighborhood with no mature foliage and plenty of construction going on. Potential buyers may be limited because many of the people wanting a custom home may prefer to simply buy a lot and build their own dream. Consider some of these issues with new developments when pricing your home:

✔ **Spec homes can set the price.** In most custom home developments, contractors are building homes for sale. These homes are called *spec homes*. If these homes match the quality of your custom home, then you're fine provided they aren't selling them at lowball prices. If they're building at less quality and lower prices, you may have difficulty recouping your home's expenses.

✔ **Too many undeveloped lots can scare buyers off.** Many resort area developments sell their land to dreamers who never build. The developers don't care because they want to sell the land for the profit. Developments with no building going on can be unattractive because most people want to live in a neighborhood. It can sometimes take decades for these developments to get production going.

✔ **You may be competing against the sales office.** If you're an early builder in your development, the sales office is still in place and working steadily on commission. Unless you're bound to list with them — which sometimes happens in new developments — the aggressive sales agents will be working hard to sell buyers on new land or spec listings held by the office. Of course they won't admit it, but they'll see any buyer interested in your finished house as a lost sale in their territory.

You need to assess these elements when deciding on the right time to market your house. If the market is hot, you probably won't have a problem. However, a lukewarm market could have you sitting for a long time or taking an unattractive price. Work with a local real estate agent to assess the marketplace. If you determine that it's not the right time to sell, then the worst is you can enjoy the new home you put so much love into.

Two-year capital gains tax implications

The government doesn't often create tax laws that benefit you, but the capital gains law is one of those rarities. (Perhaps the legislators enacted the law because they own real estate.) When your home appreciates, the money it makes is considered *capital gains* by the government. The IRS allows you to make capital gains on your primary residence tax-free with a couple of limitations. Most states follow the same rules. Each person is entitled to free capital gains up to $250,000. Here are the rules:

- ✔ You must have lived in the house two out of the last five years.
- ✔ Couples can take up to $500,000 together.
- ✔ You can only do it once in any given two-year period.
- ✔ The two-year period starts the day you move in.

With these rules, you can make a hefty amount of tax-free money simply by building homes and selling them every few years. Many of Kevin's clients have used it as a successful strategy for making tax-free income. They build a house and move in for two years or until they reach the $500,000 couple limit. When they know the property is close to its appreciation limit, the couple buys another lot and starts building. When they sell, they get all the gain on the first house tax-free. Even if it takes five years to hit the $500,000 couple limit, that still equals $100,000 a year in tax-free income. Not bad in anyone's book.

If you reach your limit on capital gains and choose to stay, you'll pay capital gains taxes on the difference at current tax rates, which can be as high as 40 percent in some states. Sit down with your accountant or tax advisor and discuss your options. Getting the facts on paper is the best way to avoid making costly emotional decisions.

Do your buyer a favor

If you do sell your house, make everything easy for the new buyer to transition in. Because he didn't go through the construction process, he doesn't know the house like you do. Here is a checklist of items to have ready for him:

- ❑ A full set of plans
- ❑ Copies of soils and engineering reports
- ❑ Copies of all inspection records
- ❑ Copies of all manuals and registrations for appliances and fixtures

- ❑ Copies of any applicable transferable warranties from the architect and contractor
- ❑ Names and addresses for the architect, contractor, and major subs
- ❑ Specific instructions on any idiosyncrasies for working features of the house

Accumulating and giving these documents to a buyer may seem like a lot of work, but having these items available can make the sale go faster and reduce the chance of late-night panic calls to your home after the homeowners have moved in.

Chapter 16

Refinancing: More Money, Cheaper Payments

You may think that after all you've been through that the financing part of your project would be finished by now. Not so fast. You aren't done yet! Just because you've already secured the financing for your custom home doesn't mean that you can file it away and forget about it — not by a long shot. In fact, you need to determine whether to keep the permanent loan that you secured after your construction financing or consider a refinance.

Most people take the path of least resistance when dealing with mortgages, which is a mistake. By taking the path of least resistance, they avoid making decisions that end up costing themselves tens of thousands of dollars. This chapter summarizes more than 20 years' expertise in helping people avoid making mistakes with their financing choices. For even more information on maximizing your financing and an in-depth explanation of the refinance process, we suggest Kevin's book, *What the Banks Won't Tell You: How to Get the Most out of Your Mortgage* (Grady Parsons Publishing).

You Still May Want Another Loan

You'd be surprised at how many people refinance their mortgages within one year of finishing their custom home's construction. Even though many construction loans come with long-term financing already in place, lives change and unexpected opportunities appear. Wanting more cash, making life changes, or simply working toward smaller monthly payments can all be excellent reasons for re-examining the financing process.

Filling the need for more cash

We're certain that your project's financials went exactly as you planned when you started your project, right? Surely you didn't have any cost overruns or added expenses or underestimations. What? You did? What a surprise!

We're going to take a wild guess that building your own home left your savings account a little less flush with cash. If so, then you may want to replenish your bank account by considering a new loan or a refinance of your current loan.

And, even if you still have some cash in the bank, you may not be finished spending yet. You probably still have landscaping costs ahead including lawns, trees, and shrubs. You may even have plans to add a deck, pool, spa, or perhaps even a gazebo. And don't forget that you'll need curtains, decorations, and, oh yes, furniture.

Some of this money may come from the sale of your existing home. But if you don't have another home to sell or any extra funds in savings, refinancing is another way to obtain the necessary cash. (Don't consider dipping into your retirement funds if you can possibly avoid it because you may have to pay any deferred taxes as well as penalties for early withdrawal.)

When you're thinking about raising cash by taking a new loan or refinancing, remember to manage your money wisely. Taking cash out to invest wisely (on improving your property, on buying stocks or securities, or on paying down credit card debt) can be wonderfully rewarding. Taking cash out to spend irresponsibly (on an expensive vacation, on a luxury car, or on a snazzy new home-entertainment system) only brings you one step closer to financial ruin — putting your new home in jeopardy.

Reconsidering how long you will stay

Another reason to consider a refinance is directly related to the *term* — that is, the duration — of the loan you received at the end of construction. If your construction loan was a single-close loan (discussed in Chapter 8), then you may or may not have had a choice in the type of loan that replaced your construction loan when your home was finished.

If the final loan was attached to the construction loan, you may find that the choice you made a year ago when starting the project was based on a whole set of assumptions that have since changed. Maybe you thought you would live in your new home forever. Or maybe you thought you would never lose your job or never get a divorce. Lives change, and you may now have the opportunity to move to a better neighborhood or take a new job. In this case, analyze the available loan programs to determine if the loan you have right now is the loan you'll need to achieve your long-term goals.

Lowering your payments

Interest rates constantly change, and new kinds of loan programs — meant to attract new customers — are always becoming available. Because you had little control over the economy when you started your project or when it ended (unless you happened to be the president of the United States or the prime minister of Great Britain), interest rates may have improved since you locked in your permanent loan. And, even if fixed rates aren't particularly attractive at any given time, the adjustable-rate mortgages and interest-only loans that we explain in the "Interest-only options" section, later in this chapter, may offer you opportunities to lower your payments significantly.

Take our advice: Now that you know what the house actually cost (ouch!) — and the real state of your finances after completion (double ouch!) — consider a change in your financing strategy that allows you to increase your personal monthly cash flow by lowering your loan payment. With the help of your accountant and a good loan officer, assessing your finances and options regularly can save you money and bring peace of mind.

Three Things You Should Know About Refinancing

Okay, so you actually need to know more than just three things about refinancing. But the three key points in this section really deserve your utmost attention. For all the rest of the information that you need to know, we highly recommend *Mortgages For Dummies* by Eric Tyson and Ray Brown (Wiley). The information that follows can give you the basic understanding and perspective necessary to make the most of your loan opportunities.

You aren't the bank's customer

You probably think that you're the bank's customer after all the money you just paid in interest and points. Guess what? That's not the way the bank sees it. Large banks actually see very little money from the interest and points on the residential loans they make, making you a small fish in a very big pond.

Here's an insider's picture of how the system works:

- ✔ After funding the loan, banks sell most mortgages to investors (the banks' true customers) within 30 days of funding (the date a check is cut to the borrower).

- ✔ Except for the rare case where banks put loans in their own investment portfolios, they make their money on the difference between the rate you pay and the rate required by the investor.

- ✔ The banks continue to manage your monthly mortgage payments — sending you monthly bills and processing your payments, and the investor pays the bank a quarter percent annually for this service called a *servicing fee.*

Because your bank handles most payments in an automated fashion, servicing a $300,000 loan earns the bank $750 at a cost of less than $50. Not bad, huh? A billion-dollar servicing portfolio can yield $2.5 million annually. The 90-percent profit margin draws the bank's customer focus.

Don't be fooled by the warm personal relationship you may have with a bank as a depositor for your personal and business accounts. For most banks, the mortgage division is completely separate. Unless you're a seven-figure depositor, you probably won't see any favoritism on your mortgage experience because the bank will likely sell your loan to an investor that doesn't know you from Adam.

Your bank mostly sees you as a necessary risk and costly annoyance required to make their servicing fees. This annoyance is one of the main reasons why many banks seem so difficult to deal with during the application and underwriting process. The banks are very well aware that they don't treat you well, but there's little financial incentive for them to do anything different. Because they see no profit from originating loans, they actually prefer to pay outside mortgage brokers more than $5,000 on a $300,000 loan just to avoid contact with you. Nothing personal, but this approach has been so successful for them that two out of three mortgages in the United States are now originated by brokers.

Your friends Fannie and Freddie

As the Great Depression wore on, President Franklin D. Roosevelt realized that the home-buying economy needed to be stimulated. The banks' collapses caused a shortage of funds to loan for home purchases. Roosevelt and the Federal U.S. government created the Federal National Mortgage Association in 1938, referred to as FNMA or Fannie Mae for short. The purpose of Fannie Mae was to buy mortgages from lenders, thereby replenishing the market with more money to loan. In 1970, Congress established the Federal Home Loan Mortgage Corporation (FHLMC) now called Freddie Mac. Both of these publicly held corporations are Government Sponsored Enterprises (GSE) with charters focused on keeping home loan money abundant and interest rates low. These institutions buy a significant number of mortgages called *conforming loans* and make financing easier and more cost effective for everyone.

The banks view new custom homes differently than existing homes

Many banks have a different set of criteria for calculating the amount you can borrow on a new custom home compared to an existing property. With a conventional refinance, the bank generally loans you money based upon the property's current market value. If you bought the house within the last year, chances are that your bank will use the sales price to determine that value. This time frame is known as *seasoning*. Banks generally frown upon the idea of letting you take out cash when your house isn't seasoned unless the *loan-to-value ratio* (the amount of the loan divided by the house's value) is significantly low enough to meet their standards.

A similar seasoning requirement may occur when refinancing a new custom home. Most banks don't consider the home seasoned until one year from issuance of its certificate of occupancy, the final project approval issued by your local building department. But, because the bank has no sales price to consider, it may require complete documentation of the home's construction costs to determine its value. Each bank uses a different formula to come up with this number; some allow for the land's market value while others use the original purchase price. Ultimately, the bank comes up with a calculation of the project's cost — representing its value — and dictates the loan amount based upon a percentage of that cost.

For example, if the total project cost based upon your documentation and receipts equals $600,000, and the bank is only willing to loan up to 75 percent of that amount, then the most you can borrow is $450,000 — even if the property may now appraise for more. If the property appraises for *less* than your cost documentation supports, however, then the bank uses the lower of the two numbers.

Rates and fees are only part of the big picture

In our many years of experience working with loans for custom home construction, we have noticed that the vast majority of people start the refinancing process by first asking about rates and fees. Don't make this mistake. Although rates and fees are an important element in the big picture, they aren't everything. In order to get the most from a mortgage, you first need to seriously assess your own unique financial needs. The loan you select is guaranteed to have a major impact — either positive or negative — on the entirety of your financial health. If you want to avoid financial problems, then be sure you pick the loan that is right for you.

Use the following questions to explore with your accountant, financial advisor, tax preparer, and loan officer before hitting the low-rate sites on the Internet:

- ✔ Do you have an appropriate amount of *liquidity* (readily available cash)?
- ✔ Are you making the right choices for your tax situation?
- ✔ Are your investments properly diversified?
- ✔ What does your cash flow really look like?
- ✔ What is your retirement strategy? (Do you have a retirement strategy?)

Your mortgage is usually connected to your largest asset, your biggest payment, and your greatest liability. To sleep well, seek the proper professional financial advice and act on it. Believe us — we've seen enough examples of people who haven't sought out the right professional financial advice to write an entire book just on bad loans. And that is one book you *don't* want to be in!

Picking the Right Mortgage Program

Before considering which mortgage program is right for you, you should have first consulted with financial planning experts and determined your financial needs as well as your long-range goals regarding your custom home and lifestyle choices. At least we hope you have. (If not, please don't pass Go or collect $200 until you do.) Okay, ready? This section includes a brief explanation of many popular loan choices available in today's market.

A mediocre loan officer can cost you big time

Most people are used to salespeople focused on giving them what they want. That's their job, right? But, with mortgages, the old saying "the customer is always right" doesn't necessarily apply. When dealing with financing, your needs are far more important than your wants. What you need is straight education. Here's a secret: Even with a so-called zero-point/zero-fee loan, *you* are paying thousands of dollars to work with a loan officer. Remember: The loan officer is responsible for bringing you her experience, knowledge, understanding, and education so that you can make intelligent, educated decisions about what is in your own best financial interest. Choose your loan provider carefully and do your own homework. A loan officer that tells you only what you want to hear — not what you *need* to hear — could very well cost you tens of thousands of dollars over the life of your loan.

Home equity lines of credit (HELOCs)

Many banks offer home equity lines of credit (HELOCs) as a *second mortgage,* that is, as a separate mortgage in addition to your primary home mortgage. HELOCs are effectively giant credit cards *secured* (guaranteed) by your home. Do you want to know the one main difference between a home equity line of credit and your credit cards? If you default on your credit cards, you get bad marks on your credit report. If you default on your loan, then the lender gets to keep your home. Some people feel that a HELOC is particularly advantageous because you can pay back the equity and you only pay interest on what you use.

Few people actually repay these loans. Most HELOCs are on a variable rate with few restrictions for only a ten-year period. If you take the money, you'll most likely have to refinance your line of credit at some point in the future. Unfortunately, we know of no way of telling today where the rates will be when that time comes. If you think you need the cash long term, our advice is to take it while the interest rates are low and lock in your rate by folding it into your first mortgage.

Adjustable rate mortgages (ARMs)

Sometimes called a *variable,* an *adjustable rate mortgage* (ARM) is a loan with an interest rate that changes on a fixed schedule. The benefit to this particular kind of mortgage is that it generally offers a less expensive payment than other loan types — at least initially. When considering ARMs as an option, the time length you plan to have your new loan as well as the current interest rate are the factors most important in making the right choice.

Several types of ARMs are available. Some adjust monthly, semiannually, annually, or at some set number of years. Some basic definition of ARM components you need to understand include

- **Start rate:** This is the initial interest rate that establishes your payment. Lower is better for you.

- **Life cap:** This is the highest interest rate your ARM can go to. Lower is better for you.

- **Index:** This is a published market rate that your loan will be tied to. If the market rate goes up, your loan payment goes up. If the market rate goes down, your loan payment goes down. Common indexes are the 11th District Cost of Funds (COFI), Monthly Treasury Average (MTA), or London InterBank Offered Rate (LIBOR).

- **Margin:** This is an arbitrary amount over the index that the lender establishes to determine the rate you'll pay. For example, your margin may be 2.25 percent over the current LIBOR rate of 1.5 percent, making your interest rate total 3.75 percent. This rate is called your *float rate*. A lower margin is better for you.

- ✔ **Rate adjustment period:** This is how often your interest rate can adjust. Less frequent adjustments are better for you when rates are going up, and better for the lender when rates are going down.

- ✔ **Periodic interest cap:** This is how high your interest rate can adjust in any one period. Lower is better for you.

- ✔ **Payment adjustment period:** This tells you how often your payment can adjust. Less frequent adjustments are better for you when rates are going up, and better for the lender when rates are going down.

- ✔ **Periodic payment cap:** This is how high your payment can adjust in any one period. Lower is better for you.

Have your loan officer explain all the details of each appropriate option. If your loan officer won't provide you with a detailed explanation, then find one who will. Get the naked truth because you deserve the right to bare ARMs. (Groan!) Seriously, lenders are coming out with new variations on ARMs constantly. Some programs — such as interest-only and short-term fixed — have great benefits and have recently become popular.

Arguments for a short-term fixed 3/1 or 5/1 ARM

3/1 and 5/1 ARMs are called short-term fixed because the interest rates stay the same for a period of three or five years before turning into an ARM that adjusts every six months or year. These loans are generally amortized over 30 years but only have a fixed rate for the first three or five years as specified. The real beauty is that they can save you as much as 1 percent in interest rate over a 30-year fixed-rate loan, without increasing your risk.

Usually, when you decide to take a short-term loan rather than a long-term loan, you're expecting a major financial change in your life within the next five years or so. Perhaps you're planning to sell your house to capture some appreciation, or maybe you're expecting a promotion or inheritance. In any case, the risk you're taking is that the prevailing market rates may increase significantly when the loan goes past its fixed period. These loans usually convert to annual adjustable-rate mortgages after their fixed terms end. On a typical $300,000 loan, the monthly payment savings can be more than $250 a month; during five years, the $15,000 in savings will more than pay the costs of a refinance.

Calculate the monthly savings on the payment of a current short-term fixed loan versus a 30-year fixed loan. Think of all the reasons you might move or refinance in the next five years. Assuming that the savings is $250 a month, ask yourself the following question: "Am I willing to bet $250 a month for five years — a total of $15,000 — that I will still have this loan in more than five years?" If the answer is no, then a short-term fixed is the right loan for you.

Banking on impounds

Some lenders offer price breaks if you accept impounds. *Impounds* are when you allow the lender to collect your property taxes and insurance as part of your monthly payment. Subsequently, your lender pays out these impounds to the appropriate insurance company and government entity. You can sometimes save as much as a quarter point upfront at closing by agreeing to allow your lender to collect these impounds. The only downside may be coming up with a bit more cash at funding to fund the impound account. You do get paid interest on the funds held by your lender, but you're not exactly going to get rich off of the small amount of interest that you'll stack up over the course of the loan.

Arguments for a negative amortization loan

The creation of the term "negative amortization" is a perfect example of a corporate bank's marketing department gone seriously wrong! *Amortization* means the spreading out of payments to pay off a loan over its time period. A negative amortization loan's interest rate fluctuates monthly. However, a low minimum payment is established, allowing the home's equity to absorb the ups and downs of the interest-rate market. Any difference between the minimum payment and the required interest rate is simply added to the loan balance. This way, your house actually helps you make the payment. Because interest rates generally rise in a hot economy, your home's value usually increases at the same time, but there are no guarantees.

The greatest advantage to this loan is that it offers you the option of paying whatever you want beyond the minimum payment each and every month. Banks often make the terms of these loans very attractive because you're sharing the risk. After five years, these loans are generally reamortized over the remaining 25 years. This reamortization may take away the payment benefits and put you in need of a refinance. This reamortization process is known as *recasting* the loan.

Examine your cash flow needs and your area's real estate market. In some areas of the country, houses go up in value so fast that the increase can equal more than your annual salary. If you're struggling for money every month, this loan lets your house shoulder that burden. Chances are that the monthly savings will have a bigger benefit to your lifestyle than the forfeited equity when you eventually sell. If you're looking for short-term cash flow solutions, the negative amortization loan is a great one.

Interest-only options

The interest-only loan is a relatively recent program offering. Normally, mortgages are *amortized,* meaning that payments are spread out equally into the future to pay off the mortgage over time. The typical 30-year fixed mortgage, for example, is amortized to pay off automatically in 30 years. Your monthly payment comprises two parts:

- ✔ The *interest due* (the cost of the loan)
- ✔ A *principal payment* (the actual money loaned to you)

Even though your payments stay level during the course of the loan, your payment amount going toward your principal becomes larger, reducing your interest and paying off the balance. For the first few years, most of your payment goes to interest (which is good news, because Uncle Sam allows you — if you live in the United States — to deduct these interest payments from your income tax).

Because most ARMs are held only for a short while, and principal reduction isn't significant, banks now offer borrowers an opportunity to have even lower payments by paying interest only. This kind of loan can provide you with significant monthly savings, without heavily impacting your long-term finances.

For example, on a fully amortized $350,000 loan at 6 percent (paying principal and interest), in the first five years you pay roughly $350 against principal each month and approximately $1,750 toward interest each month. Alternatively, you could choose an interest-only loan, saving you the $350 principal reduction each month. That money would be available to you for expenses, savings, or investment in some other asset.

So, although the interest-only loan gives you a lower payment option, you have to decide which approach makes more sense for you — tying up your money in your home (as equity) or saving $350 per month. Remember: Paying out $350 less each month can have a significant and positive impact on your cash flow whereas the amortized amount on a fully amortized loan will likely be small in proportion to your home's value. In fact, when you sell the house, you'll pay more in real estate commission than the amortized amount over a five-year period.

Zero-cost loans can be expensive

Mortgages are all about yield. A *yield* is a calculated number that takes into account both the fees that you pay and the interest rate that you get. Banks offer zero-cost loans to attract business, but raise the rate to cover their expenses.

Points and tax deduction

Because points are considered interest, they're generally tax deductible on your primary home. Some restrictions do exist, however. For example, points are deductible only on the first $1.1 million financed. With a refinance, you must amortize the deduction over the loan's term. So, if you paid $6,000 in points, you'd deduct $200 annually on a 30-year loan. If you refinance or sell the home, you can take the remaining deduction in that year. Continuing this example, in the fifth year, if you refinanced, you'd deduct the remaining $5,200 at the end of that year. Check with your accountant or financial planner to take advantage of deductibility.

Are you curious how to determine the right pricing for your loan situation? First look at when paying points may or may not be appropriate. *Points* are fees based upon a percentage of the loan amount that affect yield. One point is equal to 1 percent of the loan amount. Points are charged to lower the rate. How many points depends on the lender and the financial markets, but usually a half point lowers your rate by ⅛ percent. At this ratio, the amount of payment savings equal the points paid in roughly five to eight years, depending upon loan amount and current interest rates.

If you're taking a 30-year fixed loan, you're probably planning to have this loan for longer than five or ten years. In this case, paying points may lower your rate and payment beyond what they cost. However, if you aren't planning to own the home for more than five years, you may want to entertain the idea of a shorter-term loan such as a 3/1 or 5/1 adjustable rate mortgage (see "Arguments for a short-term fixed 3/1 or 5/1 ARM," earlier in this chapter). Having a short-term loan is a good argument for paying less points.

To obtain a zero-cost loan, the lender needs to create a *yield-spread premium (YSP)* — usually called a *rebate* — to cover the loan's points and fees. A rebate is money that the lender contributes in exchange for a higher interest rate. In order to achieve a zero-cost loan, the lender isn't going to absorb the costs of the transaction — everyone still expects to get paid. Instead, the lender is going to simply raise the interest rate to match the yield.

Before making a decision, compare which offering is best for you. Ask your loan officer to give you the interest rate with payment as well as the total points and fees for each option. Make sure you ask for all the available options. Next, pick any two options, and take the difference in total points and fees and divide it by the monthly payment difference. Doing so tells you how long it will take to make up the difference. Sometimes, the ratio of points to rate favor you instead of the bank and they can change daily. Because many of the closing costs are the same regardless of the size of your loan, a smaller loan can require a higher interest rate in order to cover the costs. You may find the difference on a long-term loan can save you thousands of dollars. It pays to ask your loan officer for all the options and run the numbers!

Pondering prepayment penalties

Prepayment penalties are common with zero-point, zero-fee loans. Such penalties protect the lender from additional losses if you pay off the loan early. The amount of these penalties varies, and the states set different legal limits. Most prepayment penalties don't exceed three years, but typically they're only for one year. A typical penalty is six month's interest on 80 percent of the balance. For example: a $300,000 loan at 6 percent interest would have a prepayment penalty of $7,200. Because the penalty is considered interest, it may be tax deductible. If you're taking a long-term loan such as 30-year fixed, then you may want to risk the prepayment penalty to save money. Some loans offer a *soft prepayment penalty;* "soft" means that you can't refinance, but you can sell the house with no penalty.

Paying Off Your Home May Be Fiscally Irresponsible

For as long as we can remember, our grandparents have been saying to us "Pay off your home, and you'll never have to worry." Of course, Grandma and Grandpa's generation also told us that cigarettes were good for us and that coffee was relaxing. Too many people today fail to take advantage of their greatest asset; they have all their cash locked up in their house where they can't get to it and where it isn't making them any money. In addition, most homeowners don't take advantage of the many tax-deduction opportunities associated with mortgages. In some cases, the government will pay as much as 40 percent of your payment so that you can take your cash and invest it in a diversified portfolio of assets with a better return.

Retirement isn't a good excuse for paying off your home mortgage. Many people are lured into a false sense of security thinking it's good not to have a house payment, but the truth is that few people die while still living at home. Most have health problems requiring special living needs, and most senior citizens end up having to sell their homes — where most of their savings actually are — to pay for healthcare or assisted-living facilities. Unfortunately, when someone is forced to sell fast, they often have to sell at a lower price. Or worse, they rack up credit card debt to get the money needed in a timely manner. Cash in hand is the only true security in the golden years. Extensively discuss these issues in this section with an experienced, qualified financial planner.

The 15-year fixed myth

An alluring loan for many people is the 15-year fixed loan. Many people believe this type of loan will protect their financial security. The truth is a 15-year loan robs you of your cash availability and interest-rate deductions. Furthermore, it locks you into a higher payment that costs you money to reduce. This loan amortizes your payments over 15 years, requiring you to put additional money every month toward your loan's principal balance. The banks don't care if you take a shorter loan or not — the reason banks offer these loans at lower rates is because they get paid back sooner and can reinvest the money. They don't want to take the time to tell you that this loan is bad for you. They don't want to pay for people to educate you. They just want to sell you what you ask for and make the quick buck.

Many people have been led to believe this loan saves them money because they're paying less interest over time. However, if you take the difference in mortgage payments between a 15-year and a 30-year loan, and you put that money into a conservative investment with a comparable yield, you would have the available cash to pay off your house in 15 years anyway. Additionally, you'd have the cash readily available to you for emergencies, as well as a better interest-rate deduction. If you work with a good financial advisor who gets you a higher rate on your investments, you'll actually have *more* money at the end of 15 years.

Also, the government allows you to deduct the interest that you're paying on your loan from your income, which reduces your tax bracket in some cases. The more you pay down the loan balance, the more you reduce the interest deduction. For individuals with a high income, mortgage interest may be the only real tax deduction that you have.

The biweekly fallacy

Many lenders offer (for a fee) a biweekly mortgage payment option, claiming that it can save you nine years of interest. Don't fall for this claim! It's just a marketing ploy!

Actually, if after reading this chapter you still decide to pay off your mortgage, you can shorten your 30-year loan by seven years without paying someone else to help you or forcing yourself into a mandatory higher payment. Most of the savings stems from the fact that there are 26 biweekly periods in a year. Because you pay half a month's payment 26 times, you're actually paying 13 monthly payments each year, which can shorten your 30-year loan in essence to a 23-year loan. Any honest loan officer can calculate accelerated payments on your loan and help you keep your money in your pocket where it belongs.

Chapter 17

Taking It Outside: The Art of Landscaping

*S*urrounding every newly built custom house is a rough construction site just waiting to be transformed into an idyllic yard. Well-planned and well-maintained landscaping adds significantly to your home's value. Using structures and plants to help shade your home can save money over your home's lifetime, too. And, believe us, nothing goes further than beautiful plants, walkways, and fences in boosting your home's curb appeal.

But don't put the cart before the horse. Take the time to first complete your home's interior before tackling outdoor projects. Here are two good reasons why:

✔ If you didn't set aside a good chunk of money to pay for *hardscaping* (walkways, fences, walls, and pools) and *softscaping* (plants, trees, and shrubs) before construction began, you may be able to refinance your newly built home to help pay for some of the costs. But you can't refinance a home that's not finished. (See Chapter 16 for refinancing advice.)

✔ Also, don't forget that the No. 1 job of those good folks building your home is getting their jobs done even if that means hauling a bunch of bricks over your freshly planted grass or tearing into your nice, neat garden to reroute the plumbing.

After you've completed your home, take comprehensive photos of its exterior and its surroundings. These photos will be invaluable tools as you purchase plant materials, talk with landscape designers, or just mull over possibilities for your yard as you sit around the kitchen table — something you'll definitely enjoy doing when your home is completed.

In this chapter, we explore the planning and installation of hardscaping and softscaping. We also discuss how to choose and work with a landscape designer. Don't forget: The more work you put into planning landscape elements, the more you and your friends and family can enjoy them.

Designing Your Dream Landscape

Think about it: You wouldn't build a home without a set of plans. So, even though just digging in and building a stone wall here, or planting a few bulbs there or a tree or two over there is tempting, take the time to devise a comprehensive landscaping plan before you forge ahead. Not only will taking the time to put together a plan save you time and money, but, believe us, it can also save you a lot of headaches, too!

The landscape plan you or a professional create offers a blueprint for your landscape's finished product. But keep in mind that you don't need to complete everything the plan specifies immediately.

Using a designer – yes or no?

Before you start drafting your landscaping plan, you need to make a major decision. Just like you chose whether or not to use an architect for your custom home (see Chapter 5), you need to decide if you want to use a landscape designer. (Take a deep breath though; a landscape designer is nowhere nearly as expensive as an architect.)

When deciding whether to use a landscape designer, you need to consider your skills, time, and budget. If you want to install items as you save money over time, you can. Or if you hope to cut costs by doing much of the work yourself, you can take your time to implement the ideas set out in the plan.

If you have the necessary tools and skills, or if you're short on cash, doing your own landscaping may be a wise investment. If, however, you're strapped for time or simply don't have the skills or proper equipment, hand the job over to a pro. What looks like a simple task, say laying a crushed stone driveway, often isn't — especially for someone who doesn't do the job for a living. If in doubt, farm it out!

Consulting the right designer

So, if you decide to use a landscape designer, you need to figure out which one can help save your home from that just-built look, instead making it look like your house and grounds landed on the earth fully matured. Actually, just to keep you on your toes, people in the landscape-design industry work under a wide variety of titles.

You may, for example, hire a landscape architect. These professionals, who often work on large projects, are bound by the strictest licensing requirements and can register with the American Society of Landscape Architects (www.asla.org).

Or, you may hire a landscape designer. Landscape designers usually have a background in horticulture, but no specific training is required to assume the title of landscape designer. Some designers complete a national certification program (and some don't), which may or may not be a big deal, depending on your particular designer's artistic sense and his ability to translate this sense into reality. You can find out more about certified professionals through the Association of Professional Landscape Designers (www.apld.com). Some landscape designers use satellite and Web technology to provide better design services. Kevin's personal favorite is www.mydesignscape.com. This company also provides excellent education on their Web site regarding the landscape design process.

When selecting a landscape designer or architect, take the extra time to do some research. Ask for at least three references and talk with previous clients to see if they're satisfied with the work before making any decisions.

Relying on other professionals

In addition to landscape architects and designers, you can call on other professionals for help, including nurserymen or growers. These helpful folks who staff nurseries or garden centers offer great experience with and knowledge of plants. For expert information on trees, look to an arborist.

Finally, the contractors who do landscape work may be licensed like other home-improvement contractors, or not, depending on your local restrictions, and they may belong to the Associated Landscape Contractors of America (www.alca.org).

No matter what type of professional you use to help create your new outdoor spaces, choose someone you communicate with well. Ask for his or her references, and check with at least three previous clients to be sure their jobs were completed in a timely manner and to their liking.

Taking the beginning steps with a designer

If you choose to hire a landscape designer, she can take your ideas, photos of the lot, and descriptions of outdoor amenities you admire to help create a landscape with the look and feel you want. (If you need inspiration, drive around several neighborhoods, noting and photographing landscape features you like.) The designer should also ask pointed questions about your lifestyle.

The designer takes note of your style preferences and puts them to work in the design she creates for you. Do you like a cottage-style garden with its lush layers of flowers and informal beds? Do you prefer something more formal and manicured? Or are you attracted to a refined Asian style?

You can't overcommunicate with your landscape designer. What is important to you? (We help you evaluate your landscaping preferences in the "Considering your wants and needs" section, later in this chapter.) Share what you want with your designer and hold her to it. Remember that your landscape design needs to complement the look of your home. Most likely, you won't be happy with a free-form cottage garden surrounding your minimalist contemporary home. Hold regular meetings with her to make sure you're both on the same page.

Revisiting your site plan

Remember that *site plan* you had drawn up way back when you purchased your land? You know, the one that shows how the house sits on your lot and how far it sits from the street and your neighbors' houses, the plan that's starting to get curled up along the edges? Time to dig it out. The site plan acts as the basis for your landscaping plan.

If you work with a garden center that has a landscape designer on staff, ask whether the garden center can deduct the fees charged for a landscape design from the cost of its plant materials you purchase. Because you have to buy the landscaping materials anyway, using your garden center's staff to draw up your plan may just be a way for you to get a lot more for your money.

The detail in your landscape plan varies based upon the designer. Some designers may refer to other drawings in your original working drawings (see Chapter 6) to reference items such as utility placement. When it's complete, your plan or plans need to show the following:

- Vegetation that already exists around your new home.
- Location of existing buildings and structures.
- Dimensions of existing buildings and structures.
- Property lines, including any irregularities such as easements granted to neighbors or lines that follow a natural feature like a river bank.

✔ Placement of windows in the finished home, to assist in planning for views from the inside (you don't want big bushes planted in front of that expensive picture window!).

✔ The site's topography — hills and valleys, where the land is low and high on the site.

✔ Site conditions, such as boggy or rocky areas. Boggy areas may be particularly noteworthy: Any place that water ponds or doesn't drain on your property may become a problem after you've built your house. Part of your landscape design needs to include grading your land away from your home, to keep standing water far from your foundation.

✔ Location of utilities, including gas lines, buried electric lines, septic tanks, sewer lines, wells, and in-ground sprinkler systems.

✔ Compass directions.

This plan will be the canvas for you or your landscape designer to paint your new home's landscape. Save it, even after work on the landscaping is done. In years to come, the plan can be very helpful, assisting you to find the location of buried utilities, for example.

Considering your wants and needs

Before you map out your new home's landscape or meet with a professional who can help you, make a list of your wants and needs for your outdoor spaces. Will you eat outside? Entertain? Meditate? Relax or play games? Do you want to spend every weekend maintaining a yard or would you rather be out on your boat or attending extracurricular activities with your kids? Is it a home that you'll only use during certain times of the year? Are you interested in creating a *xeriscape,* a type of landscaping that uses native plants and requires little or no supplemental watering? Also consider the outdoor amenities you've enjoyed at your previous homes, such as gardens, ponds, or decks, and how you can replicate those amenities around your new home.

The following sections help you consider the possibilities for your new yard.

Playing around with outdoor hobbies, games, and fun

When designing your yard, look in home catalogs or even shop the aisles of discount stores for ideas to find all the accoutrements of indoor living, moved outdoors (and even a few things that never were indoors!). You can find outdoor rugs, outdoor lamps, patio heaters, fire pits, outdoor wet bars, and pattern after pattern of outdoor serving dishes. More and more people are gathering with friends and family outside.

The land around your new home may become your own personal recreation and entertainment center. In fact, if your new place is located miles away from "civilization," you may find yourself spending more of your leisure time at home. Are there activities you want to pursue but can't right now? Want to plant a vegetable garden? Play bocce? Raise llamas? Host a neighborhood barbecue? If you have the space, you can tailor your landscape to accommodate these interests.

All types of outdoor spaces can serve as party central, and now's the time to plan for them. Will your landscape plan include a deck, patio, outdoor kitchen, or barbecue pit? Will you sip coffee on the deck in the morning? Spend afternoons around the pool? Host twilight dinners on your patio warmed by an outdoor fireplace? (If so, be sure to invite us over!)

If your family includes children or grandchildren, give thought to outdoor areas for them. A swing set is an obvious amenity, but you may also provide a children's garden, a small playhouse, a sports court for tennis, basketball, or hopscotch, or a pool.

Whether play areas are for you or for children, some preplanning can help make them perfect. How about running plumbing lines to an outdoor sink for easy cleanups after gardening? Or installing an outdoor drinking faucet to eliminate the need for kids to run in and out of the house for refreshments? If you consider what time of day you'll use the spaces, your electrical plan can include lines for proper lighting of the outdoor spaces.

In special areas set aside for children, safety dictates what materials you choose for ground cover, fencing, plant materials, and lighting. Be sure your landscape designer knows your intentions in creating play areas. Ask your nurseryman about plants to avoid near children's areas and be sure to lay down plenty of cushioning mulch or pea gravel under children's play equipment.

Factors such as your site's topography, your local climate, and what time of day you plan to use your outdoor spaces influence the kinds of amenities you select, how they're designed and built, and how you'll ultimately use them. Be sure to plan for them now, before you start building them. You'll thank us a thousand times over if you take this simple advice.

Balancing privacy and open space

Another factor to consider in planning your landscape is how the location affects your needs for privacy or open space. Your lot's size will obviously have a major effect on these needs. An urban townhouse owner, for example, may long for privacy from close neighbors, while a rural homeowner with a view of the mountains may want a greater degree of openness.

Even if your home is set deep in the woods, you may still long for a truly private "retreat" outdoors. A cozy area can be carved out in your plans and accessed by a secluded walkway. To make this area an even more special place to meditate, read, or just daydream, plan to accessorize it with items that appeal to all the senses: a bubbling fountain, wind chimes, a small pond, fragrant plants, lush flowers, a beautiful view, or a favorite piece of outdoor art.

Other outdoor spaces that may call for privacy are hot tub areas or dining areas — especially those eating spaces that you may use in the morning while relaxing in your pajamas.

Even small lots can accommodate open spaces (or create the illusion of wide-open spaces by using long sight lines). You'll need open space if you want to play games like badminton or croquet, or if you dream of stargazing at night. Mountain, valley, or water views may call for spaces free from trees. And, depending on your local climate, you may want an open space for a pool. (In hot climates, you may prefer shade near the pool. In cooler climates, you'll welcome the sunlight for its ability to warm the water.)

Dealing with your climate

Your landscape designer needs to understand your area's general climate as well your site's specific *microclimate* (the climate in your particular neighborhood). Temperature ranges, levels of annual precipitation, and the paths of prevailing winds are among the factors that dictate what types of plants and trees will thrive on your lot and those that won't (saving you the expense of replacing sick or dead greenery that isn't suited to your particular location). If you're not using a designer, find a helpful staffer at your local nursery, and ask him or her plenty of questions about which plants will work for your yard.

And, here's a bonus: Understanding these factors also can help you locate amenities such as outdoor eating areas, pools, gardens, and hot tubs so they're more enjoyable for you and your friends and family — well worth the price of admission.

Saving time and money by design

If you've ever shopped for trees, then you know just how expensive plant materials can be. But as expensive as buying plants can be, they're even more expensive when you consider the ongoing cost of watering, fertilizing, and maintenance. By choosing the right plants for your climate, you can cut down on watering and fertilizing and save yourself money in the long run. A solid landscape plan also takes into account growth patterns of trees and plants, helping you avoid expensive mistakes, like putting the shade-loving plants in the sunny parts of your yard.

You don't want to purchase, maintain, and buy fuel for a lawn mower? Or take the time to walk it around your yard every weekend? No problem. Your landscape designer can create a plan that doesn't include grass that must be trimmed every week.

And you can save on landscaping costs before you even begin construction: Proponents of natural landscaping argue that saving as much of the existing vegetation on a building site saves money in the long run and is an environmentally sensitive practice that preserves existing ecosystems. If you like the way your property looks now, then your designer can create a plan that enhances what you already have.

If you hope to protect your lot's current plants and trees (or your local municipality requires you to save certain trees), you must work closely with your contractor to prevent what Andy Wasowski (author of the eye-opening book *Building Inside Nature's Envelope* [Oxford University Press]) calls "land scraping." According to Wasowski, the practice of clearing an entire construction site down to dirt is wasteful and unnecessary. If your site beckoned to you because of its lush woods or wildflower meadow, you can (and should) create a building site without completely eliminating those features. But, you need your contractor's total cooperation. The bottom-line benefit? Every tree or plant you save during construction is one fewer you have to buy when your home is done.

Getting your plan on paper

The designer takes your ideas for your landscape and, with the help of her knowledge and experience, creates a landscape design on paper for you to review. Take the time to look over this plan carefully, asking questions about anything you don't understand. If you don't like certain elements, or some elements don't seem to mesh with your needs, tell your designer. Together, you can come up with a plan that's perfect for you.

If you're sketching out your own design, consider using one of the available software programs, such as Smart Draw (www.smartdraw.com) to help you plot out your ideas. Or, you can use the good old method of graph paper and pencil. If you draw out your site plan on paper, then top that plan with tracing paper so you can try out different landscape elements without having to redraw your site plan over and over again. Go to http://landscaping.about.com for a brief tutorial on drawing your own landscape design. (See Figure 17-1 for an example of a landscape plan.)

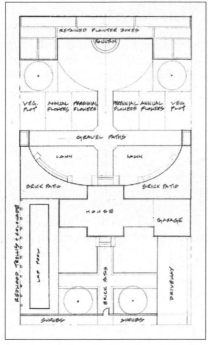

Figure 17-1:
An example
of a
landscape
plan.

Courtesy of Tecta Associates Architects, San Francisco

Putting Your Plan in Action — Hardscaping

You can't hang draperies in your new home until you have finished putting in your windows and walls. It's the same outside: You must wait to install plants in your yard until you've created the structure of your landscape. These structural, or nonliving, elements are referred to as hardscaping.

As with most of the elements of your new home, you can build your hard-scape features or hire someone to do it for you. If you have the skills, tools, and, most of all, time to tackle these projects, go for it. Otherwise, you can find the professional contractors needed (masons for concrete or stone work, carpenters for wood fences or decks, and so on) by asking your general contractor, family, friends, neighbors, or landscape designer if you're using one for recommendations. Be sure these contractors are licensed and bonded according to your local codes and ask for and check their references.

Planning a patio, Daddy-o

Materials for patios are the same as those for walkways and driveways (see Chapter 14). The material you choose depends on your style preference, climate, and budget. To create a patio that you'll use often and spend little time maintaining, make sure you find the best possible place for it in your landscape and install it correctly.

The material you choose to pave the patio will be its *floor*. You can also plan for "walls" for the patio — which may consist of just an ankle-high border of flowers or a half wall of stacked stones. If you want a bit of shade for the patio, you can build some type of overhead shelter. It could be a lightweight, open trellis, a more substantial kind of trellis that rests on pillars (also called a *pergola*), an awning attached to an adjacent house wall, or a permanent roof. Because you'll probably eat and drink on the patio, make sure it's easily accessible to your home's kitchen or an outdoor refrigerator and grill.

Patios, obviously, are located at ground level. If your home plan calls for an outdoor space that is raised (either a little or a lot), choose a deck. (See the next section, "Lounging on the deck," for more information on decks.) If the ground around your home slopes radically, a patio may not be an option. When adding a patio, make sure you don't impede the flow of rainwater away from your home.

Lounging on the deck

If you were to take an aerial snapshot of just about any housing development in the nation, you'd see decks sprouting off the back of every home. Homes just don't seem complete anymore without some form of deck or outdoor living space. The deck you plan can be a simple square or rectangle, or a multilevel affair that terraces out over your landscape. It can bridge the space between your home and your pool, offer space to eat, relax, and look up at the stars, boast a hot tub for soaking, or accommodate a snazzy gas grill.

With so many books, Web sites, and magazines showing photos and plans of decks, gathering ideas for yours needn't be difficult. We talk about decks that are attached to the house in Chapter 14. You may decide to add an unattached deck away from the house and pay for it after the house is complete. Be careful not to end up with a deck in a place that's so sunny or so shaded that using the deck will be unpleasant.

Your choice for deck materials includes

- **Cedar,** which is also naturally resistant to decay and insects.
- **Other exotic woods,** such as ipe, which may offer the benefit of more closely matching the look of the flooring inside your home.

- **Manufactured wood,** a durable and almost maintenance-free product made from wood flour and plastic resins. Plastic wood offers the added benefit of recycling unwanted materials (such as sawdust and pop bottles) into something usable.

- **Pressure-treated pine,** which is widely available and inexpensive.

- **Redwood,** which naturally resists decay and insects, but is quite expensive and not always available at local home-improvement stores.

Choose a material based on your budget, your tolerance for maintenance tasks, and the look you want to achieve. You can use these same materials to build other outdoor wood structures, such as gazebos, children's play sets, benches, and trellises. If you're interested in recycling, like the idea of a low-maintenance material, and don't find the look of manufactured wood offensive, why not choose it as a sensible alternative to other woods? In a few years, you'll be happy to be sitting on your deck sipping a cool drink while your neighbors work up a sweat sanding and staining their decks.

The art of fencing

When you or your landscape designer are putting together your landscaping plans, remember that the uses for a fence are endless. They can keep wild critters out of your yard and hold your pet critters in. They can define gardens or seating areas. A tall fence brings privacy and sometimes shade. Some fences are only for show, others to keep peace between neighbors. All provide structure in the landscape — a kind of frame for the masterpiece you paint with plants and trees.

Fencing materials run the gamut from wood to stone to metal to plastics. Which material you choose depends on

- Your budget (wood is less expensive than stone or metal)

- Your preferred degree of maintenance (wood requires much more maintenance than stone or metal)

- Whether you want to build the fence yourself (digging those postholes is a lot of work!)

- Your neighborhood's codes and restrictions (which may govern fence height, materials, or setback from property lines)

- The function of the fence (are you looking for privacy or decoration?)

Before you start digging the postholes for your new fence, be sure you know (and mark!) the exact location of your lot lines and utility lines. You don't want to have to move the fence because it sits on your neighbor's property. And heaven knows you don't want to dig and hit your water line. Also be sure the fence you have planned adheres to your local building codes and any restrictions your neighborhood or community spells out.

Patio stone can be Gorges

In 1992, the Chinese government decided to flood the Yangtze River Valley with a super dam that will be the largest on the planet. After the dam in the Three Gorges area of China is complete and the valley is flooded, 1,600 villages will cease to exist, except as a natural habitat for fish under 600 feet of water. What does this have to do with hardscaping? Plenty! One very enterprising business couple has made arrangements to take apart the ancient buildings, stairs, and roads piece by piece so that you can enjoy the beauty of this ancient stone in your yard or patio. A piece of history can be yours for a mere $45 per square foot. For more information, check out www.rhodes.org and see how entrepreneurs Richard and Pamela Rhodes are single handedly saving history.

Building great walls

Walk around the ruins of an abandoned house and you'll likely see stone foundation walls, a stone chimney, or maybe even stone walls still standing long after the rest of the home has disintegrated. Stone walls are simply built to last.

The enduring strength of stone walls — and the skill required to build them — accounts for their high price tag. Still, nothing adds charm or substance to a home's gardens and landscape quite like a stone wall.

If you want a true mortared stone wall outside your home, you need to hire a skilled mason. To find potential candidates, ask your general contractor, the staff at your local garden center, or other homeowners who have masonry walls you admire.

Building a stone wall yourself requires time and patience — and no small amount of skill. Whether you do it yourself or hire it out, another way to save money is to substitute bricks or manufactured stones. Today's manufactured stones very closely resemble their natural counterparts and are available in a wide range of colors and styles. If you need assistance, ask for information at your local garden center or check out *Landscaping For Dummies* by Phillip Giroux, Bob Beckstrom, Lance Wallheim, and the editors of the National Gardening Association (Wiley). You may also find that boulder walls, which must be installed with the help of large machinery, can hold back large amounts of earth.

If you've made the decision that nothing but stone will do for you, the place to look for the lowest prices is the stone yard or quarry. Many garden centers and home-improvement stores carry small amounts of rock and gravel, but you'll find it more economical to buy bulk quantities at a stone yard. Buy enough, and the stone yard will likely deliver and unload it for free, if you're within a reasonable distance.

Adding water

Even if you can't live on the coast or near a lake, you can still create a water view in your landscape. If you want to be lulled by the sound of water, but want a low-maintenance water feature, choose a small, portable fountain. For a larger investment in time or money, you can create a pond complete with koi fish, water lilies, and a waterfall. You can buy a pond kit at your friendly, local home-improvement center or garden pond specialist, or buy plastic sheet pond liners and even preformed plastic ponds. The choice is yours.

But, don't let the fish have all the fun — you can enjoy a swim at home, too, if you plan a pool for your yard. Today, pool design is light years ahead of the rusty metal above-ground pools you may remember from a few decades back. What type of pool you choose — from a lap pool to a hot tub to a shallow wading pool — depends on what you want to use it for, from physical activity and relaxation to water therapy and entertainment.

As with many outdoor amenities, the type of pool you ultimately choose will be dictated by your budget, willingness to perform maintenance tasks (or pay to have them done for you), local climate, and community codes and restrictions. Remember: A pool is (just about) forever. After you have it, you have it, and if you don't keep up with maintenance, you'll have little more than a mosquito-and-algae-infested swamp on your hands. Still, in some neighborhoods, a pool is almost a necessity for resale. Your local climate and customs affect your decision to add a pool or not to add a pool.

Lighting the way

At this point, take advantage of the groundwork you laid for your landscape lighting when you designed your home's lighting plan. Ideally, you had your electrical contractor add a few electrical sockets and light sockets to the house's exterior. (See Chapter 14 for more information.) With the proper wiring in place, you can create a wide variety of lighting effects outside your home, which can make your home more beautiful and safer.

Consider these options, but remember, sometimes less is more. You don't want your outdoor lights keeping the neighbors awake all night.

- **Uplighting** creates interesting shadows by placing a light at the base of a structure or tree that shines up.
- **Downlighting** is sometimes referred to as moonlighting because it creates light from above. The effect is achieved by placing a light fixture high in a tree.

- ✔ **Feature lighting** makes certain objects the stars in your outdoor show. You may choose to spotlight a favorite tree, a backyard pond or fountain, your doorway, an outdoor art object, or another architectural feature of your new home.

- ✔ **Grazing** consists of installing a light close to an interesting surface, such as a brick or stone wall or the bark of a tree. The light shines across the surface, showing off its texture.

- ✔ **Low-voltage lighting** fixtures are available in a wide variety of styles to match your home's looks. These lights add romance and safety as they reveal the boundaries of a yard or line a walkway.

- ✔ **Utility lighting** comes into play when you need bright light in an outdoor area, such as near the grill, by a set of steps, or near the garage where you'll be unloading your car.

Love to light your home at the holidays? Talk to your landscape designer about how to make the process easy. She can give you some suggestions on placing outlets, camouflaging cords, and hanging fixtures.

Leafing Out — Softscaping with Plants

Your frame is complete and now you can begin to paint broad brushes of color into your landscape and add layers of leafy texture with plants. Before you can begin, however, take a good look at the soil around your new home. Fill dirt and construction waste brought to the site may have rendered much of the soil around your home less than desirable.

Get advice from your landscaper on amending the soil to make it more hospitable to new plants and trees. Then take the time to implement the measures she suggests. You're wasting time and money and killing or crippling valuable plants if you stick them in crummy soil.

Planting trees after construction

The first living things to arrive on your site will probably be trees. If you've done a careful job of protecting trees during construction, as we suggest in the section "Saving time and money by design," earlier in this chapter, you may not need to plant new ones. Keep in mind, however, that saving trees isn't an easy task. See Chapter 5 for more information.

As you plant new trees, keep their growth patterns in mind and how their placement may affect your views from inside. Will the new evergreens you plant bring unwanted shade year-round to your new home's sunroom? Will you be able to see the beautiful blossoms of your new cherry tree from your kitchen window?

Remember that trees grow underground, too. Root systems can damage foundation walls and underground sewage lines. A good landscape plan eliminates these potential problems by protecting the trees' root systems from damage that can be inflicted by heavy trucks parking too close to a tree, or excavation that disturbs the roots.

As you choose your trees, keep in mind your love of (or distaste for) yard work. Do you like to rake? Remember that trees that drop their leaves mean extra yard work. Trees with blossoms or berries planted near your deck or front walk may bring clean-up tasks with them. Some trees require little pruning, others a lot. Ask questions when you're designing your plan or when purchasing trees.

Adding texture with shrubs

Between the tree and the flowering plant, you have the shrubs. These versatile plants offer color (sometimes year-round), provide form, and bridge the gaps between low-growing flowers and taller trees. What type of shrubs you and/or your landscape designer choose depends on where they are located, what you want in terms of leaf shape and color, and how much space you can give them.

If you want a low-maintenance landscape, don't choose shrubbery that requires constant trimming. You can also keep weeding to a minimum and help plantings retain moisture by adding landscape coverings such as wood chips, wood mulch, or crushed rocks.

Bloom time — Annuals and perennials

For many people, flowers are the icing on the landscape cake. But most garden experts advise you that flowers are short-lived. When choosing flowering plants for your yard, keep the plants' other characteristics in mind. What are the color, size, and texture of the leaves? How tall will the plants grow? Are they invasive?

Fortunately, you can easily get answers to these kinds of questions from the multitude of gardening books available, your landscape designer, or even by looking at neighbors' yards. The latter method may be the most helpful if you plan to build your home in a region that's new to you. You may not be familiar with the types of flowering plants that thrive in the climate. Taking a tour of local gardens can give you a quick tutorial.

Saving money on plants

Poking around your neighbors' gardens (with their permission, of course) can actually save you money. Your local gardeners and your landscape professional may agree that certain types of flowering plants just don't do well in the climate, or require a high degree of maintenance. You can lower the cost of your landscape by choosing plants that require little watering and less fertilizing in your climate.

If you're lucky, you may score a few free plants during those visits to green-thumb neighbors.

Many plants must be divided seasonally to maintain their shape or their health. Offer to help your neighbors thin their stock, and ask if you can take divisions home to plant. Other gardeners' castoffs could be a windfall for the barren plains around your new home.

You may also find good deals on plants and shrubs at Lowe's or Home Depot. But remember to stick to your plan when buying plants. Don't just buy something because it's a bargain — it may not thrive in your garden.

For quick answers to basic questions, along with in-depth information on landscaping and gardening, visit the National Gardening Association's Web site at www.nationalgardening.com. Or check your local bookstores for *Gardening For Dummies,* 2nd Edition, by Mike MacCaskey and Bill Marken, and *Landscaping For Dummies* by Phillip Giroux, Bob Beckstrom, Lance Walheim, and the editors of the National Gardening Association (both by Wiley).

Part V
The Part of Tens

In this part . . .

Like every handy *For Dummies* book, this part provides several quick reference chapters to help you along in the process. These chapters are great for generating ideas or solving problems. We cover potential problems, solutions, and mistakes. We also give you cost savers and ideas to make your project environmentally friendly.

Chapter 18

Ten Common Custom Home Mistakes and How to Avoid 'Em

In This Chapter
- ▶ Steering clear of design mistakes
- ▶ Dodging financing mistakes

Sure, it'd be great if every custom home project went off without a hitch, but in the real world, you're going to have at least a few hurdles to overcome. When you're risking your life's savings on something as important as your home, you can't afford to head the wrong direction too many times. In this chapter we examine ten of the most common and potentially damaging mistakes and suggest ways of avoiding them before they happen.

Designing a House Unlike Others in the Neighborhood

Some people may assume that a major reason you're building a custom home is because you can't find a house you like built by anyone else. However, presuming you can build anything you want wherever you want is a mistake. If you have an unlimited supply of money, then the future value of your home may be unimportant. But for most people, their home's value is as much security for them as their other savings.

Building a house significantly larger or smaller than the other houses in the neighborhood can make your house undesirable and harder to sell in the future. You can avoid this problem by asking your real estate agent to show you houses that are selling in the neighborhood, and you can also ask for information on houses that have sold within the last three to six months. You don't have to design a house exactly like your neighbors' homes, but you can safely assume that future buyers looking in this neighborhood will be looking for similar houses. However, you may face a challenge if you're building in a new subdivision where homes haven't yet been built or sold. You may have to rely on neighboring subdivision information. Check with the homeowners' association (HOA) to find out what other homes are being built. Talk with the other landowners if you can to find out their plans. Many HOAs have minimum and maximum square footage guidelines exactly for this reason. Chapter 5 addresses the design process more extensively.

Attempting to Build outside the Design Review Guidelines

The design review process can be long and difficult enough without battling for an exception to the established building guidelines. Neighborhoods and cities set up these guidelines because most people like conformity in their neighborhoods. You may in the end win a prolonged battle with the design review committee, but it will cost you time and money (and perhaps the goodwill of your new neighbors).

Review the design guidelines before you buy the property. Design review committees are under no obligation to make exceptions. If you're going to buy in a development of custom homes, then accept the fact that you'll need to conform to the neighborhood. The more you design within the guidelines, the faster and easier your approval will be. See Chapter 6 for more information on the design review process.

Thinking the House Will Be Worth What It Costs

Many people just assume that if you spend $500,000 to build a house, it'll be worth $500,000. Some of the truly optimistic may think that if you spend $500,000 to build a new house, it'll be worth $1 million. Nothing could be further from the truth. A house is worth the price that someone is willing to pay for it. Research the houses that are selling in the neighborhood to see what amenities people are paying for. Just because you spent $10,000 apiece for ten antique Italian doors doesn't mean that prospective buyers will pay $100,000 more for your house than the house next door.

As soon as you understand the value of the neighborhood, you can create a budget that fits your situation. If you're going to finance your project, you're going to be dependent on a budget that fits in line with the appraised value. You can find more information on the lending and appraisal process in Chapters 8 and 9.

Paying Off the Lot Before Starting to Build

Paying off your lot before you start to build continues to be the single most common mistake that consumers make. Although banks previously required a paid-off piece of land before they would provide construction financing, this practice has been dead for more than a decade.

Today, banks are concerned with your *liquidity,* or how much ready cash you have available. After good credit, money in the bank is the next most important criteria for loan approval. Putting your money in your land ties up your money where you can't get to it and the bank can't verify it.

The best way to avoid this mistake is to finance your land with a low down payment and hoard your cash until your construction loan is finished. You'll have plenty of opportunities to put that money into the house after your house is completed and you have the final tally on your project's cost. In the meantime, cash in your hand can save you from a variety of trials and tribulations along the path to your dream home. See Chapter 3 for information on land financing and Chapter 8 on the lender's approach to liquidity.

Waiting for Permits to Investigate Construction Financing

Unfortunately, no standard order for the entire process of designing and building a custom home exists. Logic can dictate some of the steps but sometimes logic can be deceiving. Even though the construction loan doesn't take place until you're ready to build, you need to keep in mind several considerations in the design process — such as the compatibility of the house in the neighborhood and making the budget fit a lender's guidelines — that can impact your ability to get financed. Even the wrong approach to land financing can delay your project for years. By the time you have submitted your permits, making significant changes to appease lending guidelines may be too difficult.

The earlier you engage someone knowledgeable in construction lending the better. Not all lenders will spend the time, especially if you're a few years away from building, but don't let that discourage you. Take a loan officer to a free lunch. Some brokers work hand-in-hand with you for several years to make sure that every decision you make will be consistent with getting the project completed in a financially palatable manner. To familiarize yourself with what lenders are looking for, be sure to check out Chapter 9.

Applying for the Construction Loan Too Early

Applying for a construction loan too early may sound strange considering we tell you not to take too much time in the previous section. The key difference is the issue of *applying* for the construction loan. Get as much information as you can on how the lender works and how to make your project fit its lending criterion. However, applying for the loan more than two months before you're ready to break ground can be a waste of time and money. Furthermore, you may disclose information that was unnecessary or may change in your favor. Some information could make getting the loan at all difficult.

The safest general rule is to wait until your final plans are ready to be submitted to the building department. (We lay out this process in Chapter 6.) Applying at this time generally puts you within 60 days of breaking ground. The lender won't fund the loan without permits and because most of your documentation, such as the appraisal and credit report, is good for 90 days, you'll have a little cushion. If it takes longer, don't panic. You can renew or restart the loan process, possibly with some minor additional fees. Try and get as close as you can to the 60-day mark and communicate openly with your loan officer on your timing.

Applying to Too Many Lenders

The construction lending community is a small one. In fact, the number of national lenders that offer all-in-one construction and permanent loans (see Chapter 8) number less than ten. Many banks and brokers submit their loans to these lenders. If your loan gets submitted to the same lender through two different sources and the information is inconsistent, you may be turned down completely. If that lender is the only one that fits your project needs, you may find yourself without any loan at all.

Make sure your loan officer tells you who will be funding your loan. Loyalty goes a long way with good construction loan officers, so do your shopping first and then start the application process. After you start, make sure you keep aware of how the loan process is moving along. If you know of or anticipate any hitches, try to resolve them with your loan officer. If you aren't getting cooperation, move your loan. You're under no obligation to stay with anyone who doesn't take care of your needs.

Borrowing the Minimum to Get By

The only way to know exactly what a custom home project will cost is to finish it. Therefore, the only time you can know your minimum loan needs is after completion. We can think of nothing worse than running out of money in a custom home project. Don't be a penny pincher. The amount of money you save by taking a loan for $100,000 less is only $1,000 to $2,000 of tax-deductible money — a relatively cheap insurance policy that protects you from the tens of thousands of dollars it could cost you on an underfunded project.

Kevin's company has the motto "If someone offers you money, you should take it." You can always reconcile the proper loan amount after you have moved in and paid for everything in the project.

Spending Your Own Money First

Some people decide that they'll start the building process with their own savings and then get a construction loan when they've used up all their funds. This theory has some major problems:

- The banks need to see cash reserves in your bank accounts (see Chapter 9) and won't count any of the money you have already put into the project toward the funds.

- If you've been using credit cards to fund construction, your credit scores decrease, causing even bigger problems for qualification.

- Your project can come to a grinding halt in the middle if you run into any financing difficulties, which can cost you even more money and may force you to take a more expensive loan to bail out the project.

Have your financing all set to go when you're ready to break ground. You can apply for a loan after you've started, but try to avoid those unnecessary risks if at all possible.

Selling Your Existing Home Before Building the New One

If you're building in the same region where you currently live, we can't think of many reasons to sell your existing home until you're ready to move into the new one. Here are a few good reasons for keeping the old home during the build:

- Many lenders don't factor your old house in their calculations for qualifying.

- You can take the necessary cash with a refinance or credit line at favorable rates and payments. Doing so gets you the cash today without having to sell the house. (See Chapter 8 for more details.)

- You can avoid the cost and hassle of moving twice.

The custom home process has enough challenges, so the less disruption you can make to family life until the house is finished, the better!

Chapter 19

Ten Great Ways to Lower Construction Costs

In This Chapter
▶ Saving money without cutting corners
▶ Keeping your project within your budget

Nearly every would-be homeowner has a budget for his or her new home. Your budget may be small or it may be "Bill Gates" big, but either way, you'll feel better about your home in the long run if it stays within your financial expectations.

As you read through these tips, you can see that staying within your budget often goes hand-in-hand with a smooth construction process — one that causes you fewer sleepless nights, helps maintain a civil relationship with your general contractor (and your significant others!), and gets you moved in to your new home faster.

An added bonus to keeping your construction costs down is that you'll still have enough money when the home is finished to buy furniture, rugs, and accessories. When the urge to splurge arises during the design or construction of your home, focus on the end result. When your house is finished, do you want to have a new house that's devoid of furniture or do you want a cozy, fully furnished home?

Build the House That Suits the Land

Do you want to eat up your construction budget fast? Try to build a home with a walkout basement on a lot with no slope. You'll waste a wad of money excavating and pushing dirt around. A better idea? Plan your home to fit the site you have.

If you don't have much level ground, for instance, design a home with a small *footprint* (the amount of space the building occupies on the ground) that doesn't spread out for thousands of square feet on the first floor. Or if you do have a sloped site, you can include that walkout basement in your plans. You can get livable space for less by taking advantage of your site's characteristics.

A little bit of forethought goes a long way, of course. If you want to have open space for a pool, a vegetable garden, or a pasture, look for land that isn't completely covered with trees. The costs to prepare your site will be lower, and you'll be ready to build sooner.

Make It Tall to Keep Your Budget Small

Time for a quick comparison: Take a 3,000-square-foot ranch-style home and a three-story home that provides 1,000 square feet on each floor. Which one do you think offers more size for your cents? In the first home, you have to dig and build a 3,000-square-foot foundation. In the second home, the foundation measures only 1,000 square feet. In the first home, the roof must cover 3,000 square feet. In the second, the roof's surface area is significantly smaller. Plumbing lines and ducts may need to run much longer distances in the one-level ranch, adding to material costs and possibly affecting the efficiency of plumbing, heating, and cooling systems.

Using this oversimplified example, you soon can realize that building a home "up" is usually less expensive in terms of materials and site-preparation costs than building a home "out."

Of course, your lifestyle or your community restrictions may require a one-story home. But if you have any flexibility, consider stacking your living spaces to save money.

Keep It Simple and Tasteful

To cut down on construction costs, simplify, simplify, simplify. Take the time during the design phase to look at your plans with a critical eye (see Chapter 5). Do you really need a home with three-story turrets flanking the front door, for instance? These rounded structures will add time and money in construction costs, and could create a home that looks overdone and out-of-place with its neighbors.

Opting for a simple roofline (think of the classic lines of Colonial homes) instead of a complex roofline (like the multitude of peaks and valleys in today's McMansions) is a wise cost-cutting move. You can save money during construction on labor and materials and possibly save in the long run when it's time to have the roof replaced.

Watch out for luxury amenities that can quickly break the bank. Put two whirlpool tubs in your home, and your budget will begin to bubble up, too. You'll need to buy the tubs, increase the size of the bathrooms, and install a second water heater just to fill them (not to speak of the maintenance costs over the years as expensive pumps burn out and require periodic replacement).

To help cut costs, avoid anything curved, including ceilings, walls, windows, and stairways. Keeping everything simple brings an added bonus: Creating a home with classic lines, traditional good looks, and common-sense amenities means it should retain its charms year after year and provide good resale potential.

Use Design Elements to Eliminate Costly Materials

Sometimes getting the look you love doesn't have to break the bank. Granite countertops, for instance, are an upscale choice for today's kitchens. If your budget doesn't allow for granite, however, you can choose a less expensive material, such as ceramic tile, and mimic the look of granite.

Another tried-and-true design trick is to keep the ceiling in your entryway low and then have it lead into a living space with a higher ceiling height. The entry will feel intimate and warm, and the change in ceiling height will make the living space feel dramatic. What's more, you don't have to worry about spending money to furnish or light a towering entry hall, and you don't lose square footage on the second floor.

As you design your home, you'll discover that choices you make affect your budget. For example, say you choose an open floor plan: It could save you money that would normally be spent framing out interior walls and purchasing and installing interior doors, but it may cost more to frame the structure of a large, open space. The key is to review your design and construction estimates carefully and ask your designer and contractor plenty of questions about costs.

Choose and Order Items in Advance

Picture yourself going through the process of designing your home. You've chosen every light fixture, floor covering, doorknob, and window style for every room. You've picked one dishwasher from among hundreds of models. You've finally decided on granite for the countertops. You've hit the wall.

Now, your architect or contractor wants you to choose a tile for your kitchen's *backsplash* (the vertical wall area between the countertop and the bottom edge of the overhead cabinets). The mere thought of spending another afternoon looking at samples spikes your blood pressure. "I don't care about the tile," you say through clenched teeth. "Just put in whatever tile is standard."

The contractor specifies a builder's standard tile and provides his cost estimate for your kitchen. All is well until a few months later when you spot a beautiful hand-painted tile while on vacation. You know the tile would be perfect for your backsplash. You tell the contractor where to order it, and suddenly the line item for tile on your budget is blown.

What's the solution to going over budget on item after item? Choose and order as many fixtures and finishes as possible in advance. When a contractor or architect is forced to estimate or put in an allowance for an item in your budget, the budget loses clarity. Put in a couple dozen allowances and your cost estimates are really just ballpark figures — not a rock-solid price you can count on.

Use Surplus Materials for an Old-Style Home

"If you love old houses," the saying goes, "build one." Following the lead of classic homes can give your new dwelling character and may help you cut construction costs. Architectural salvage firms are a great resource to find vintage doors, knobs, ironwork, mantelpieces, and drawer pulls. Adventurous homeowners seeking a more eclectic look also may have good luck finding mismatched door and drawer hardware at flea markets or yard sales.

Be a smart shopper, though. Buying old light fixtures may cost more in the long run if you need to rewire them. Old, primitive windows have plenty of character but lack the energy efficiency of new window units. If you find an old window you love, use it in an interior wall, instead. Light can still shine through it between rooms, and you'll still be able to enjoy its craftsmanship.

Group Simple Windows for a Custom Look

Think you can't live without a three-story, arched-top window in your new great room overlooking the golf course? Or perhaps a bay window (one that juts out from the exterior wall) for your breakfast area is on your must-have list. When you need to pinch pennies, you may be able to create these custom window looks with standard, less-expensive window units.

You and your architect or designer can experiment with stacking rectangular windows of various sizes to create an interesting pattern in your great room that offers timeless charm. Looking ahead, you may be saving yourself the hassle of covering round-shaped windows with window treatments, too.

To "fake" a bay window, you can just group two or three windows side by side. You'll get almost as much light, without having the expense of framing out the protruding bay window. You can still have a window seat built in under the flat windows, or simply place a bench up against that wall when your home is finished. You'll even be able to spring for a bevy of beautiful throw pillows for the bench with the money you save on the windows.

Avoid Change Orders

Yes, changing your mind is your prerogative. But when building your own home, exercising that prerogative may cost you a pile of dough. Every contractor can tell horror stories about the house that got out of hand. During construction, a homeowner decides to add a swimming pool, put in another bathroom, move a wall, or expand a window. Some of these requests may actually be quite simple in design or concept, but they all have ramifications — sometimes major — in terms of time and money.

Most builders or general contractors charge for every change order you request during construction. In addition to the change-order fee, your alteration could cost more in materials or labor. It may also set your construction timetable back, stretching out your construction loan and increasing the amount of interest you pay on that loan. The best way to save money during construction is to stick to the plan you carefully created for the home during the design phase.

The exception to the rule is when a design error becomes readily apparent during the building phase. You realize a door blocks a hallway if it's left open, for instance, or a recessed light can't be installed in a ceiling because a duct is in the way. Obviously, these mistakes require that changes be made to the original drawings. That's why your construction loan officer wanted you to create a contingency fund in your building budget. (See Chapters 8 and 9 for more information about contingency funds.) You want your fund to cover the little emergencies that pop up during construction.

If possible, visit your site daily during construction to help you spot potential design flaws quickly. If you see something that doesn't work, tell your general contractor — not the subcontractor's crew. Your daily visits may also provide the added benefit of letting subs know you care about the house and appreciate their best work.

Go Faux

You don't have to be a wealthy patron to have artful walls in your new home. With today's products and tools for homeowners, figuring out how to apply faux finish techniques is easier than ever. The results are walls that look like masterpieces but don't cost a year's salary.

Start your faux finishing project with a trip to your local paint store. The staff can recommend techniques to create the look you have in mind — from Tuscan adobe to washes of tropical pastels. You can add a big splash of character for a small amount of cash. For example, before you tackle your dining room walls, buy a small amount of the needed paints or glazes and practice your technique on a scrap of drywall. If your practice run goes as planned, then you can invest in all the supplies you need.

Books and magazines offer pages of inspiration for faux finishes. Two books in particular that can light your way are *Decorative Paint and Faux Finishes* by Linda Selden and Carol Spier (Sunset Books) and *The Paint Effects Bible: 100 Recipes for Faux Finishes* by Kerry Skinner (Firefly Books, Ltd.).

Get Creative

The key to creativity is to keep an open mind. Remember, just because the homes you see as you tour local open houses all have formal dining rooms doesn't mean you have to spend money to build and furnish a room that you may never use.

Don't become so enamored of some high-end finish that you can't see the benefits of a lower-priced but perfectly suited alternative. Who knows? Your choice to use painted wood bead-board to cover your bathroom walls rather than tile may make you a trendsetter in your neighborhood.

Brainstorm with your family, architect, contractor, and spouse — together you can come up with ways to cut costs without looking like you've cut corners.

Chapter 20

Ten Common Stuck-in-the-Middle Problems and Their Fixes

*W*hen building your custom home, the question isn't whether problems will occur during the project, but when and how you will handle them. In this chapter, we list the most common problems and give you some solutions to help.

My Home Is Behind Schedule

Even the most thought-out schedules are estimates at best. Mother Nature and other difficulties, such as late materials and labor shortages, can put a building project behind schedule in a hurry. Weekly communication with your contractor can keep you apprised of the progress. Make your first question at every meeting "Are we running on time?" If the answer is no, then discuss options such as rescheduling subcontractors.

If your project is encountering a significant delay, that holdup can often affect the availability of the subs. (For example, the plumber can't exactly install the kitchen faucets if the carpenter hasn't installed the cabinetry.) Discuss the situation with your contractor. Communicate your concerns and financial risks. Talk about workable solutions, such as the option of using alternative subs. Doing so may help you get the project done on time. The risk is that the contractor may not have as good of a relationship with the new sub, and the project could suffer in the quality of workmanship.

If your contractor isn't responsive to your needs, then you need to have a heart-to-heart conversation with him and review your contract's terms. If the contractor is being negligent per the contract's terms, you may need to discuss changing contractors. However, taking this step should be a last-resort issue at best. (See Chapter 11 for more detailed information on contracts and remedies.)

Ultimately, you don't want to rush the project. As long as you keep in mind the cost of any lender penalties (discussed in the following section as well as in Chapter 10), a couple of extra months isn't worth cutting corners on a house that will last lifetimes.

My Construction Loan Is Expiring and the House Isn't Complete

Don't bury your head in the sand now if your construction loan is about to expire and your house isn't complete. You need to remain calm and resolute. Most construction loans carry some sort of penalty for going over the allotted time limit. Although these penalties can be steep, they often can be negotiable.

After the project is underway and you detect a problem, start the conversation with your lender as soon as possible. Because your lender may charge you a smaller penalty for one single extension, estimate a time of completion and negotiate a specific time for completion. After you're past your deadline, taking it month by month may make your lender concerned, and your lender will be more likely to charge you the full penalty amounts.

Discuss with your lender the minimum work required to roll the construction loan into a permanent loan so you can finish as soon as possible. Often you may be able to put some work off (such as landscaping and finishing work) until after the construction loan is rolled into permanent financing as discussed in Chapter 15.

My Contractor Wants More Money in a Fixed-Price Contract

A *fixed-price contract* exists when the contractor has given you a set price and is responsible for all the expenses with the remainder being his profit. You can read more about this in Chapter 11. When a fixed-price custom home project goes over budget, it's either because the contractor's estimate was too low or because the consumer keeps making changes to the original plans.

If you find yourself in this position, carefully review the terms of your contract and discuss options with your contractor. The solution depends on what blew the budget:

- **The contractor made mistakes.** If your contractor underestimated the project, sit down with her to renegotiate. Don't play the blame game right away. Get to the root of the problem and discuss how the contractor can assure you of resolving those problems in the future.

  If the contractor is in breach of contract due to her mistakes, it still may be in your best interest to negotiate a fee for finishing that you can both live with. If she's losing money, she may cut corners where you can't see them or, worse, file bankruptcy and leave you hanging with a half-finished house. Some people consider involving attorneys in these negotiations. Be careful; attorneys can add intimidation and fear to an already stressed environment and can easily escalate the situation into an adversarial disaster. It also may be cheaper to pay your contractor than to cover the attorney's fees.

- **You made more change requests than your contract allowed.** Your requests for changes in the original plans are called *change orders* and they are *your* issue. (See Chapter 11 for more on change orders.) The contractor can't absorb the cost of you changing the project every week. Fortunately, you have direct control with change orders. Consistent communication can prevent sticker shock at the end. Discuss the time and cost of each change before the contractor begins the process. If the contractor has already made the changes and you find yourself looking at the bill you didn't expect, you're probably on the hook, and you'll need to cover these changes out of your own pocket.

The bottom line is you want to keep the project moving no matter what the issue. Often, delays that concern costs end up causing more problems and expenses than the actual cost increases themselves.

Everything Costs More Than My Budget in a Time-and-Materials Contract

A *time-and-materials contract* exists when you're responsible for all the material and labor expenses, and the contractor makes his money as an additional percentage of the total cost. (You can read more about this contract type in Chapter 11.) As you can imagine, "everything costing more than I thought," is the most common complaint in the custom home–building world. Estimating a construction project can be very challenging when you consider the huge list of materials and labor involved. Even the most experienced contractors can't account for all the variables involved.

When costs are your financial responsibility, the best way to deal with cost overruns is to keep a close watch on them. Weekly communication with your contractor can keep the overruns from getting out of hand. If you do notice an overrun problem, you can discuss options on reducing expenses later in the project if possible. Having access to extra cash can help solve the immediate problem when it arises. For more about the need for cash, read Chapter 7.

The Bank Won't Fund the Draws

Usually you encounter the problem of the bank not funding the *draws,* or requests for money, when the percentage of your house that has been completed isn't relative to the percentage of funds you have taken from the bank. In other words, if you have requested 50 percent of the money and the bank says the house is 40 percent complete, you may have difficulty getting the bank to cough up the other 10 percent. We explain the fund disbursement process in more detail in Chapter 10.

Again, communication is key. The bank will ultimately fund you all of any section, such as foundation or framing (see Chapters 12 and 13), when complete, but you have to get the sections complete first. Arrange for a phone meeting with the bank and contractor. Make sure that everyone is in agreement about what stage the project is in. If you can't pry any money loose from the bank, make arrangements with your contractor to get to the next payment stage. If you have cash or credit available, you should be able to easily resolve the issue by paying for things yourself until you and the bank agree on the percentage of completion. In the final stages, incomplete paperwork, such as lien releases explained in Chapter 11, can cause delays, so be sure to keep your records readily available as we advise in Chapter 2.

I'm Out of Cash

Running out of cash is a dangerous problem to have in a construction project. It usually stems from a combination of poor planning and putting too much cash into the project upfront. If you do run out of cash, try the following possible resources for cash. However, understand that these resources also have their drawbacks if you're attempting to solve a long-term cost overrun issue, as we discuss in the previous sections in this chapter, rather than just a temporary cash flow issue.

- **Credit cards:** You can reimburse yourself after the project pays off, so don't be shy, *charge it!*
- **Friends and family:** You're borrowing a short-term loan, and who are you going to be entertaining in your new home? Let them share in all the joys of custom home ownership.

✔ **Home equity loans:** If you haven't already tapped the equity in your current personal residence, now is the time. You'll likely sell the house when finished anyway.

✔ **Retirement accounts, such as 401k or stock/margin accounts:** Borrow if you can, liquidate only if you have to as a last resort. *Margin accounts* are credit lines secured against stock accounts. Because they're secured, the interest rates are generally low, but you could risk having to repay them in a hurry if your stock shares devalue. Liquidation of any retirement accounts could result in penalties and taxes, so be sure to consult your accountant and financial advisor before taking action.

Remember that your first priority is to keep the project moving. Otherwise you risk losing all your investment and your credit. As soon as the house is finished, you have many options on how to repay the cash you borrowed, as we discuss in Chapter 16.

The House Is Having Some Engineering Problems

As soon as the build has started, enduring engineering problems related to technical, structural design is always a tough situation. Luckily, most engineering issues are relatively small, or they're caught in the plan check process we discuss in Chapter 6. If you have a major issue, such as a foundation that won't support the weight or a structural wall that needs to be moved, you need to immediately communicate with the architect, contractor, and lender to determine a plan to resolve it. Often the solution requires some form of redesign and new approval of the changed plans with the building department.

Most often a minor engineering issue will simply create a delay or expense. If you start to have several small issues come up, taking the plans to another engineer for a second opinion may be worthwhile to go through the plans with a fine-tooth comb. This way you can get all the potential problems cleared at one time and move on with the project. Too many separate delays and cost increases can take their toll and create other problems that we address in this chapter.

My Contractor and I Can't Get Along

Sadly the problem of a contractor and a homeowner not getting along isn't as uncommon as you may think. Face it, in life, making friends and hiring employees is hard enough. When you selected a contractor, you picked someone with whom you thought you could trust your life's savings and who exhibited good craftsmanship and excellent management skills. Furthermore, you selected someone who you hoped you were compatible with in your work methods in stressful situations. Finding someone who met all those requirements was a big challenge.

Work hard to put personal emotions aside. Always remember that you and the contractor have the same objective of getting the house finished on time and on budget in a quality manner. Be willing to sacrifice personal ego and correctness for getting to the objective. The contractor won't be in your life after the house is finished, but you'll live with the home (and the emotional experience) that got you there for years to come.

My Contractor Walked Off the Job

This problem can translate to time and money if you don't act quickly. Unless your contractor left for reasons unrelated to your project, don't attempt to reengage him. Chances are if he walked out on you, the communication was past the point of repair. You can reconcile the money and breach of contract issues later as we outline in Chapter 11. Depending upon your lender, you may be able to take over the helm and act as your own general contractor. If that option isn't possible, then you need to work with the lender to find a contractor soon. Try the contractors you interviewed before to see if they have availability.

If the house is fairly far along in construction, you may be able to find a smaller remodeling contractor to finish the job. Although getting bids will be important, you don't have the option to be too picky about price. Look for a contractor that can step in and do quality work. Of course you need to investigate the contractor in the same basic ways we suggest in Chapter 2. You may need to compromise on issues such as price and timeline to get the house back on track for completion.

This Project Is Breaking Up My Marriage

Building a custom home can take its toll on family and relationships. Keep your cool. Find ways to communicate and make the project fun to work on. Kevin has been giving this advice to his clients for more than a decade and only one of the hundreds of projects he has funded ever ended up in divorce.

If things do get tough, focus more attention on your relationships and let your contractor take over more of the day-to-day operations and worry on your custom home project. If things get really tough, then by all means seek out marriage counseling before you get past the point of no return.

Chapter 21

Ten Helpful Custom Home Resources

*Y*ou can never have enough extra resources when embarking on a custom home project. In this chapter we give you a few handy resources for finding land and arranging your financing as well as listing the major associations that work with custom home people. We also list some of our favorite specialty resources for owner-builders, log homes, and environmental help.

Land Sources

Lotfinders (`www.lotfinders.com`) is an Internet resource focused specifically on helping the consumer find land to build a custom home. The Web site focuses on education and then on listing resources and instructions for hunting down properties. Lotfinders works with developers to publicize properties for sale. They also help prospective homebuilders to identify landowners with property not listed. They have direct resources for lot financing as well.

Other resources for finding land also on the Internet include

✔ `www.Realtor.com`: This site can get you to any land listed in the Multiple Listing Service.

✔ `www.land.net`: This site has a large database of land for sale as well, although much of it is commercial land or *raw land,* which is land that's not prepared for building, which we define further in Chapter 3.

Stratford Financial Services

No company has been more dedicated to real estate financing education than Stratford Financial Services (www.customhomeexperts.com). In its 15-year history, the company has financed thousands of homes and more than 800 custom home projects. Its Web site has tons of custom home articles and essays designed to keep consumers from making costly mistakes when financing real estate. The site offers expert information on custom home financing, purchase money, refinancing, as well as groundbreaking information about owning rental properties.

Founded by Kevin in 1989, Stratford is also a one-stop resource for lot and construction financing. As a broker specializing in construction loans, the company can assess your situation and help you determine which lenders have the best programs for you. The company's expert staff can keep you from making costly mistakes by going direct, as we outline in Chapters 8 and 9.

Log and Timber Frame Homes

Two publishing companies serve as the best all-around resources for log and timber home enthusiasts. Our favorite is Goodman Media, publisher of *Log Homes Illustrated* magazine and *Timber Homes Illustrated* magazine (no, they don't have a swimsuit edition). Log on www.loghomesillustrated.com for more information. Janice is the editor of *Timber Homes Illustrated,* so you know you're getting solid custom home help. *Log Homes Illustrated* is committed to bringing Kevin's insightful information to you each and every month to help you with the most challenging part of custom home building. This company also produces a large number of log home shows nationwide under the name Log Home Expo. These shows are a must for the log home newbie. You can also find the schedule at the Web site.

The other company that provides resources for log and timber home enthusiasts is Log Home Authority (www.loghomeauthority.com), which has two major magazines that service the log home industry. *Log Home Living* and *Log Home Design Ideas* are excellent supplements for the log home information hound. This company is the other major producer of log shows.

If you're like most loggies, you can't get enough information and pictures about log homes, so there's no reason not to subscribe to all four magazines in this section. Often the different perspectives can help you determine what's right for you.

Finished Plans

Since 1907, The Garlinghouse Company has served as America's premier publisher of home plan books and project plans. Its Web site (`www.garling house.com`) is a great online resource. Garlinghouse provides great information for converting bought plans into approved plans. Hanley Wood is also a good resource. You can find Hanley Wood's plans online at `www.eplans.com` and `www.dreamhomesource.com`.

Not only can you pick from hundreds of thousands of plans to buy online, but you can also peruse local bookstores for plan magazines such as *Good Housekeeping Home Plans* from Hanley Wood and *1295 Best-Selling Home Plans* published by Garlinghouse. Even if you don't buy the plans, the magazines can be excellent resources to help you explore what you like and don't like.

Hanley Wood also publishes a number of trade magazines for the building industry. If you're interested in seeing things from your contractor's perspective, check out *Builder* magazine or *Custom Home* magazine. These magazines are also great resources for suppliers. If you don't see these magazines in the bookstore, you can check out the selection at `www.hanleywood.com`.

Home Depot Expo

When you're in the market for appliances, fixtures, and cabinetry ideas, there is no better showcase open to the public than Home Depot Expo. This store exhibits an amazing designer showcase where you can see how bathrooms and kitchens come together. Whether or not you buy anything really isn't important. What is important is that you can check out different designs to see what your likes and dislikes are. (For example, you have a ton of decisions to make on fixtures, as we show you in Chapter 5. A day at Expo with a pencil and digital camera can save you evenings on the Internet and billable hours with your architect.) Log on to `www.expo.com` to find a design center near you.

American Institute of Architects (AIA)

If you want to know all about architects, the American Institute of Architects (AIA) is the place to start. You can go to the AIA Web site (`www.aia.org`) to read up on the subject of architects and the public. You can also search the architect-finder service by location. The AIA has local offices in most major cities, which can be a great resource for meeting architects in person.

National Association of Home Builders (NAHB)

At first glance, the National Association of Home Builders (NAHB) Web site (www.nahb.org) looks to be for builders only. It deserves a closer look. You can find plenty of consumer information about builders and the building process spread throughout the site. The site has detailed explanations of the building process and listings of trade shows that you can attend to see building materials and processes. Kevin's favorite is the Pacific Coast Builders Conference (PCBC) held in San Francisco every June.

Also connected with NAHB is the Custom Home Builders Council, which focuses on the issues important to the custom home-building industry. For owner-builders (see the next section) or those of you with large complicated projects, you may find the NAHB a useful peer group.

One other resource is www.homebuilder.com, which is a consumer site connected with NAHB. It's a decent source for information and houses a referral source for finding contractors.

For Owner-Builders

If you're considering being an owner-builder, which we discuss briefly in Chapter 1, you can utilize two strong resources for owner-builders not quite looking to do it all themselves. U-Build-It (www.ubuildit.com) is a set of franchised dealers that educate and consult with you through the entire construction process. They become your advisor and advocate at each stage. Many of their franchisees are former builders and contractors. They do charge fees but are generally less than hiring a contractor.

The other resource, Complete Owner Builder Services (COBS; www.cobs homes.com), is a one-stop shop for owner-builders. Based in San Diego, COBS combines its expertise along with supplier relationships, such as Home Depot, to help you save money and achieve your goals in the owner-builder process. The COBS approach is to start with the finances, which we certainly approve, but you aren't obligated to use its lender.

Building Green

Building Green provides authoritative, independent information on environmentally responsible design and construction. You can gain insight and information to share with your builder. Building Green's resources include

- ✔ **Environmental Building News (EBN):** A newsletter featuring comprehensive in-depth features, product reviews, news, opinions, and more, on a wide range of sustainable building topics

- ✔ **GreenSpec Directory:** Featuring more than 1,850 green building products carefully screened by the editors of *EBN,* plus guideline specification language

- ✔ **BuildingGreen Suite:** Providing instant online access to *EBN* and thousands of past articles, cross-linked with the *GreenSpec* product database and high-performance building case studies

To find more information, visit www.BuildingGreen.com or call 800-861-0954. Check out more green ideas and resources in Chapter 22.

Wild Ones Native Plant Landscaping

Started in Milwaukee in 1977, this 50-chapter organization promotes environmentally sound landscaping practices incorporating the use of native plants in landscaping for houses and buildings. If you're interested in native plant landscaping, check out Wild Ones at www.for-wild.org.

Why go native? Because native plants evolved and adapted to local conditions over thousands of years, so they're more vigorous and hardy. Also, they can more easily survive harsh weather conditions. After native plants take root, they require little or no irrigation or fertilization. They're more pest and disease resistant. Native plants are perfect for low-maintenance gardening and landscaping. And they bring original beauty back to regional areas. Need we say more?

Chapter 22

Ten Ways to Make Your Home Green

*N*o matter what color your new home will be on the outside, you can make it green inside. The best way to make your home *green,* or environmentally friendly, is to design it that way to start. Look for ways to reuse, recycle, and reduce as you design and build. Talk with your contractor, architect, or designer about your desires for a green home. With any luck, these building professionals can rise to the challenge. If your requests meet with blank stares or eye rolling, consider choosing new team members.

Keep in mind that each of the tips in this chapter is merely a door into a much larger topic. If one of these ideas sparks your interest, utilize the helpful resources provided to find more information.

Making the Best Use of Your Lot

If you haven't purchased your land yet, you have a great opportunity to create a home that makes the earth happy. Look for a site that requires minimal excavation and earth moving. Choose a spot that won't expose your home to extreme elements.

After you find the perfect spot, work with your designer or architect to put your home in the best possible place on the site. For general resources and information on siting your home, go to www.buildinggreen.com and read Chapter 5.

Avoiding an Oversized House

You've scrimped and saved for your new house. Now you're ready to blow all your money on a home that can only be described as massive. Remember that building a huge home is going to make a big impression — and not just on your neighbors. Every square foot you put into your home will cost you and, in the long run, the environment. In general, bigger homes require

- ✔ More disruption of land during construction
- ✔ More energy to heat and cool
- ✔ More maintenance and cleaning
- ✔ More raw materials to build

To help keep your square footage down to earth, take a long hard look at your true needs and desires during the design phase (see Chapter 5). Do you really need a two-story foyer in your new home? Or would that space be better used as an extra closet or an office area on the second floor? A good clearinghouse for information on keeping home sizes reasonable is the Web site www.notsobighouse.com, a spin-off of the book *The Not So Big House: A Blueprint for the Way We Really Live* by architect Sarah Susanka (Taunton Press).

Planning for the Ages

You can make your home green by keeping the future in mind as you build. Making good long-term choices can result in environmentally friendly results. For instance, imagine that you're nearing the end of your construction project and running short on cash. You decide to carpet your upstairs bedrooms instead of putting down hardwood because carpet is less expensive, and you figure you can always replace the carpeting in a few years with something more durable. The yards of inexpensive carpet you install can give off unhealthy fumes and quickly look shoddy. When you remove that carpet, it goes straight to the landfill.

For information on environmentally friendly products, log on to www.oikos.com.

Making Your Landscape Earth-Friendly

Your home's landscape can do its part for the environment, too. Trees, especially, are worth their weight in gold. (To see the many benefits trees offer, check out the National Arbor Day Foundation's Web site at www.arbor day.org.)

If you live in a region with four seasons, choose deciduous trees, such as maples and oaks, that can shade your home in the hottest months, and then lose their leaves and allow sun to warm your home in the winter. If your new home will be in an area that stays warm year-round, look to evergreen trees and shrubs that offer continuous shade, especially for your home's sunny southern and western exposures.

Choosing low-maintenance shrubs and plants that require little fertilization and watering is also kinder to the environment. Reducing the amount of lawn, which has to be mowed, fertilized, and watered, is a significant step forward in keeping your home green.

Your landscape can be friendly to other living things, too. Many people enjoy planting flowers and trees that attract butterflies and birds, and help sustain other small animals. For information on creating a backyard wildlife habitat, go to www.nwf.org/backyardwildlifehabitat.

Using Alternative Energy Sources

Do you remember the time when only back-to-the-land hippies heated their homes and water with solar power? Now, with today's technological innovations, harnessing the sun's power to make your home more livable and energy efficient is much easier. Keep in mind that, in general, installing a solar system costs more than tapping into a public electric utility, but you'll enjoy the benefits of a nonpolluting, quiet, and renewable source of energy. And, if you build your new home in an area that's far away from the public utility grid, solar or alternative energy systems may cost less than running excessively long utility lines. Be aware, however, as we state in Chapter 9, that most lenders don't make loans on houses that are not on the public power grid.

If you want to ease into solar power, look to *passive* solar heating for your home. A passive system allows for the home to absorb the sun's warmth during the day, and then release the warmth back into the home during the evening. To work, the system may only require a bank of south-facing windows located near a *heat sink*. This heat sink is typically a mass of masonry, such as a tall stone chimney or a thick concrete floor, that the sun gently warms during daylight hours. The masonry naturally radiates the warmth through the nearby space. (A supplemental heating system may be needed to warm other parts of the house.)

If you're building in a warm or hot climate, you don't want to inadvertently create a passive solar heating situation that will make you uncomfortably warm inside. As you design your home, avoid large banks of windows with southern or western exposure that can channel the sun's heat into your home, forcing you to waste energy on cooling.

With a bit more investment (and the blessing of your local building codes and neighborhood restrictions), you can create a solar heating system with *photovoltaic panels* (devices that convert the energy of sunlight into electric energy) on your property or your home's roof. Newer panels are smaller than ever — some roof shingles today incorporate photovoltaic cells. With a solar power system, you can create energy for your home using a truly renewable resource. The cost of these systems varies from region to region. For more information, talk to a local solar energy contractor or visit www.mysolar. com. The Rocky Mountain Institute also offers informative briefs on solar power, which you can download at www.rmi.org.

Of course, the sun isn't the only alternative source of energy. You can also capture the power of wind or water to make your home more livable. Each of these alternative energy systems requires certain conditions. For example, a hydro-powered system needs a body of water, of course, and a wind-powered system only works on a site that offers dependable and accessible air movement. So if you plan to use one of these systems, keep their requirements in mind as you look for land to purchase.

Talk to your architect or designer or speak with an alternative energy professional to find out more information. You can also find tons of info on the Internet. A few places to start include www.realgoods.com and www.solar livinginstitute.com.

Going On an Energy Diet

A great way to save energy is to simply use less of it. Plan ahead to create a home that sips energy instead of guzzles it.

Talk with your architect, lighting designer, or general contractor about a whole-house lighting plan. A whole-house system can allow you to program your home's light fixtures to go on or off at certain times, based on your family's needs, and can allow you to turn off lights throughout the home with a single switch. A master switch means you're less likely to leave lights on when they're no longer needed.

Another way to use less energy is to reach for the stars — Energy Stars, that is. The government's Energy Star program identifies products designed and built to be energy efficient. Choose windows, appliances, fixtures, and even building products for your home that have earned a thumbs-up from the program.

You can find out more about the federal government's Energy Star program by visiting www.energystar.gov. For more information on building an energy-efficient, weather-tight home, go to www.rmi.org.

Cutting the Fumes

Help the earth and your family by choosing paints, stains, and finishes with low volatile organic compounds (VOCs). These materials give off fewer harmful gases and fumes and make your home more pleasant and healthy.

Most paint stores now offer low VOC paints. You need to ask cabinet makers and furniture makers about the possibility of using low VOC finishes on their products. Talk to your architect or contractor about ways to reduce VOCs and other harmful gases such as carbon monoxide in your home.

For tips on maintaining healthy air quality in your home, go to the American Lung Association's Web site at www.lungusa.org.

Making Recycling Easy

If you consider recycling your family's glass, aluminum, and plastic a hassle, more than likely you don't do it. However, keeping your home green by separating your trash isn't that difficult, and you can do your small part for the environment.

Build a recycling center in your new kitchen or garage and plan for a place to store cans, bottles, paper, and containers. For example, one ingenious idea employed recently involved creating a trio of chutes cut through a kitchen wall and into an adjacent garage. Each chute emptied into a tub. With this setup, recycling simply involved tossing a can or bottle down the appropriate chute and carrying the full tubs in the garage to the curbside for pickup. If the chutes are too much for you (or if cutting chutes in your garage wall would violate fire code requirements in your area), make sure you save a little extra space in your kitchen, pantry, or garage to store simple recycling receptacles.

Some rural communities don't offer recycling pickup. If recycling is important to you, ask about pickup services or drop-off centers when you're purchasing your land.

The National Recycling Coalition's Web site offers more information about recycling. Go to www.nrc-recycle.org.

Using What You Have

Recycling and reducing are great ideas, but so is reusing. As you go through the design process, consider how you can furnish some rooms with pieces you already own. Some may work as is, others may need reupholstering; but in the long run, you can save money and cut down on the resources required for creating new furnishings.

Selecting salvaged architectural items, such as interior doors, mantels, bricks, and doorknobs, is another good way to keep items out of the landfill and give some charm to your new home. Keep in mind that although old salvaged windows will give your home character, primitive window units probably won't provide the kind of energy efficiency that makes good environmental sense. For example, if you find a vintage window you love, use it in an interior, partition wall, or have its glass replaced with mirrors and hang it over your mantel or in your entryway.

You can find architectural salvage companies in your area by looking through your local phone book. Other companies that specialize in salvaging old house parts advertise in specialized periodicals such as *Old House Journal*. The magazine's Web site is www.oldhousejournal.com.

Stashing and Storing

Growing your own food is a beneficial choice for the environment. With a successful garden, you can cut trips to the grocery store and eat fresher, healthier foods.

If a vegetable garden is in the plan for your new home, be sure you also create space to store your harvest. Earth-sheltered pantries and root cellars can keep food cool and dry. For information about building these storage areas, go to www.greenhomebuilding.com.

Think you have a black thumb and could never grow your own food? You can still plan ahead to decrease the number of trips to the grocery by creating pantries and storage areas in your home for provisions bought in bulk. When designing your home, build a pantry or two in your kitchen. Also consider adding outbuildings that could offer cool, dry, and critter-free storage for food, beverages, or other supplies.

Index

Franklin Planner, 26
Freddie Mac, 290
French drain system, 223
frost line, 228
full-documentation loans, 142
fumes, reducing, 347
fun things to do during construction, 40
funding and underwriting fee, 158
furnace. *See* HVAC
furnishing, cash reserves required for, 127
future expansion, affecting design, 99

• *G* •

garage, 103
gardening, 348
Gardening For Dummies (MacCaskey, Marken), 316
gas, connecting to, 227, 242
general contractor. *See* contractor
girders, 235
Giroux, Phillip (*Landscaping For Dummies*), 312, 316
good faith estimate (GFE), 55, 152–153
grading, 116, 221–222, 305
green homes. *See* environmentally friendly homes
GreenSpec Directory, 341
grout, 272
gutters, 256–257

• *H* •

Hamilton, Gene (*Bathroom Remodeling For Dummies*), 101–102
Hamilton, Katie (*Bathroom Remodeling For Dummies*), 101–102
handcrafters of logs, 69–71
handrails, 105
hard costs, 176–177, 192
hard (private) money, 60, 147, 166
hardi board (Hardie Board), 252–253
hardscaping, 309–314
hardware, 104, 105, 106, 269–270
header, 236
header joists, 235

hearths, 259–260
heat sink, 81, 345
heating, ventilating, and air conditioning. *See* HVAC
height restrictions for property, 50, 65, 117
HELOC (Home Equity Line Of Credit), 148, 293
HOA (homeowner's association), 117
home. *See* custom home; existing home
Home Buying For Dummies (Tyson, Brown), 45
Home Decorating For Dummies (McMillan, McMillan), 284
Home Depot Expo, 106, 339
Home Equity Line Of Credit (HELOC), 148, 293
home equity loan, 335
home magazines, 64
homeowner's association (HOA), 117
homeowner's insurance, 280
hot tubs, 313
Hurley, Pat (*Smart Homes For Dummies*), 107
HVAC (heating, ventilating, and air conditioning)
  choosing, 105
  energy-efficient, 107
  installing, 243–244
  mechanical plans for, 113
  solar heating, 81, 345
  subfloor access to, 236
  ventilation, 102, 107, 342
hybrid home, 74

• *I* •

IBC (International Building Code), 113
I-joists, 233
Ikea, 106
impounds, 295
income
  debt-to-income (DTI) ratio, 57, 166, 170–171
  lender's requirements for, 169–170
  leveraging home as, 134–137
  no-income-qualifier loans, 142
  "stated income" loans, 57, 58, 142

• S •

salary. *See* income
SBC (Standard Building Code), 113
seal (caulk), 235
seasoning, 172, 291
second home, 165
second mortgage, 293
second story. *See* stories, number of
secured debt, 129
Selden, Linda (*Decorative Paint and Faux Finishes*), 330
self-funded projects, 185–186
seller carry back, 42, 54
selling custom home, 284–286
selling existing home, 148, 281
septic engineer, 14
septic system, 47, 116–117, 226
servicing fee, 290
setbacks, 48–50, 218, 219
sewer, 47, 226, 241–242
sheathing, 247–248
sheet metal roofing, 255
shingles, types of, 255–256
short-term fixed ARM, 294
shrubs, planting, 315
sidewalks, requirements for, 47
siding, 252–253
sill, 235
single-close construction loan, 140–141
SIPs (structural insulated panels), 72, 73
site plan, 111, 304–305
size of home
  assessing land value based on, 50–52
  CC&Rs affecting, 65, 118
  conforming to neighborhood, 52–53
  environmentally friendly, 344
  requirements for, 95–96
Skinner, Kerry (*The Paint Effects Bible: 100 Recipes for Faux Finishes*), 330
slate roofing, 255, 256
Smart Draw program, 308
Smart Homes For Dummies (Briere, Hurley), 107
soft costs, 126, 176, 192
softscaping elements of landscape, 314–316
software for designing floor plans, 90
soils engineer, 14

solar heating, 81, 345
solar power, 106, 345–346
sole plate, 236
soundproofing, 249
source of funds for recent cash, 172
spec (speculative) home, 165–166, 285
Spier, Carol (*Decorative Paint and Faux Finishes*), 330
spindles, 265
square footage, 29
stairs, 78, 95, 237, 265
Standard Building Code (SBC), 113
"stated income" loans, 57, 58, 142
stated income, stated asset loan, 142
stated income, verified asset loan, 142
steel framing, 67–68
stick framing, 237
stock brokers, 132
stocks, 132–133, 172
stone exterior, 253
storage space, 98
stories, number of
  cost related to, 326
  footprint and setbacks affecting, 50
  framing for second story, 234, 235, 237
  height restrictions affecting, 65
  zoning restrictions affecting, 95
story poles, 119
Stratford Financial Services, 146, 338
straw bale homes, 79
street lighting, requirements for, 47
stretcher strip for carpeting, 272
stringer joists, 235
structural insulated panels (SIPs), 72, 73
structural plans, 113
stucco, 253
studs, types of, 67–68
style of home, 64–67, 117. *See also* construction techniques; designing home
subcontractors
  finding, 21
  list of, 211–212
  paying, 194–195, 197, 209–210
  role of, 15
  working with, 21, 39, 212
subfloor, 234, 235–236
sun, affecting orientation of home, 91–92
suppliers, 15, 20, 197, 210–211

BUSINESS, CAREERS & PERSONAL FINANCE

0-7645-5307-0

0-7645-5331-3 *†

Also available:
- Accounting For Dummies †
  0-7645-5314-3
- Business Plans Kit For Dummies †
  0-7645-5365-8
- Cover Letters For Dummies
  0-7645-5224-4
- Frugal Living For Dummies
  0-7645-5403-4
- Leadership For Dummies
  0-7645-5176-0
- Managing For Dummies
  0-7645-1771-6

- Marketing For Dummies
  0-7645-5600-2
- Personal Finance For Dummies *
  0-7645-2590-5
- Project Management For Dummies
  0-7645-5283-X
- Resumes For Dummies †
  0-7645-5471-9
- Selling For Dummies
  0-7645-5363-1
- Small Business Kit For Dummies *†
  0-7645-5093-4

HOME & BUSINESS COMPUTER BASICS

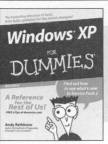

0-7645-4074-2

0-7645-3758-X

Also available:
- ACT! 6 For Dummies
  0-7645-2645-6
- iLife '04 All-in-One Desk Reference
  For Dummies
  0-7645-7347-0
- iPAQ For Dummies
  0-7645-6769-1
- Mac OS X Panther Timesaving
  Techniques For Dummies
  0-7645-5812-9
- Macs For Dummies
  0-7645-5656-8

- Microsoft Money 2004 For Dummies
  0-7645-4195-1
- Office 2003 All-in-One Desk Reference
  For Dummies
  0-7645-3883-7
- Outlook 2003 For Dummies
  0-7645-3759-8
- PCs For Dummies
  0-7645-4074-2
- TiVo For Dummies
  0-7645-6923-6
- Upgrading and Fixing PCs For Dummies
  0-7645-1665-5
- Windows XP Timesaving Techniques
  For Dummies
  0-7645-3748-2

FOOD, HOME, GARDEN, HOBBIES, MUSIC & PETS

0-7645-5295-3

0-7645-5232-5

Also available:
- Bass Guitar For Dummies
  0-7645-2487-9
- Diabetes Cookbook For Dummies
  0-7645-5230-9
- Gardening For Dummies *
  0-7645-5130-2
- Guitar For Dummies
  0-7645-5106-X
- Holiday Decorating For Dummies
  0-7645-2570-0
- Home Improvement All-in-One
  For Dummies
  0-7645-5680-0

- Knitting For Dummies
  0-7645-5395-X
- Piano For Dummies
  0-7645-5105-1
- Puppies For Dummies
  0-7645-5255-4
- Scrapbooking For Dummies
  0-7645-7208-3
- Senior Dogs For Dummies
  0-7645-5818-8
- Singing For Dummies
  0-7645-2475-5
- 30-Minute Meals For Dummies
  0-7645-2589-1

INTERNET & DIGITAL MEDIA

0-7645-1664-7

0-7645-6924-4

Also available:
- 2005 Online Shopping Directory
  For Dummies
  0-7645-7495-7
- CD & DVD Recording For Dummies
  0-7645-5956-7
- eBay For Dummies
  0-7645-5654-1
- Fighting Spam For Dummies
  0-7645-5965-6
- Genealogy Online For Dummies
  0-7645-5964-8
- Google For Dummies
  0-7645-4420-9

- Home Recording For Musicians
  For Dummies
  0-7645-1634-5
- The Internet For Dummies
  0-7645-4173-0
- iPod & iTunes For Dummies
  0-7645-7772-7
- Preventing Identity Theft For Dummies
  0-7645-7336-5
- Pro Tools All-in-One Desk Reference
  For Dummies
  0-7645-5714-9
- Roxio Easy Media Creator For Dummies
  0-7645-7131-1

SPORTS, FITNESS, PARENTING, RELIGION & SPIRITUALITY

0-7645-5146-9

0-7645-5418-2

Also available:
- Adoption For Dummies
  0-7645-5488-3
- Basketball For Dummies
  0-7645-5248-1
- The Bible For Dummies
  0-7645-5296-1
- Buddhism For Dummies
  0-7645-5359-3
- Catholicism For Dummies
  0-7645-5391-7
- Hockey For Dummies
  0-7645-5228-7

- Judaism For Dummies
  0-7645-5299-6
- Martial Arts For Dummies
  0-7645-5358-5
- Pilates For Dummies
  0-7645-5397-6
- Religion For Dummies
  0-7645-5264-3
- Teaching Kids to Read For Dummies
  0-7645-4043-2
- Weight Training For Dummies
  0-7645-5168-X
- Yoga For Dummies
  0-7645-5117-5

TRAVEL

0-7645-5438-7

0-7645-5453-0

Also available:
- Alaska For Dummies
  0-7645-1761-9
- Arizona For Dummies
  0-7645-6938-4
- Cancún and the Yucatán For Dummies
  0-7645-2437-2
- Cruise Vacations For Dummies
  0-7645-6941-4
- Europe For Dummies
  0-7645-5456-5
- Ireland For Dummies
  0-7645-5455-7

- Las Vegas For Dummies
  0-7645-5448-4
- London For Dummies
  0-7645-4277-X
- New York City For Dummies
  0-7645-6945-7
- Paris For Dummies
  0-7645-5494-8
- RV Vacations For Dummies
  0-7645-5443-3
- Walt Disney World & Orlando For Dummies
  0-7645-6943-0

GRAPHICS, DESIGN & WEB DEVELOPMENT

0-7645-4345-8

0-7645-5589-8

Also available:
- Adobe Acrobat 6 PDF For Dummies
  0-7645-3760-1
- Building a Web Site For Dummies
  0-7645-7144-3
- Dreamweaver MX 2004 For Dummies
  0-7645-4342-3
- FrontPage 2003 For Dummies
  0-7645-3882-9
- HTML 4 For Dummies
  0-7645-1995-6
- Illustrator CS For Dummies
  0-7645-4084-X

- Macromedia Flash MX 2004 For Dummies
  0-7645-4358-X
- Photoshop 7 All-in-One Desk Reference For Dummies
  0-7645-1667-1
- Photoshop CS Timesaving Techniques For Dummies
  0-7645-6782-9
- PHP 5 For Dummies
  0-7645-4166-8
- PowerPoint 2003 For Dummies
  0-7645-3908-6
- QuarkXPress 6 For Dummies
  0-7645-2593-X

NETWORKING, SECURITY, PROGRAMMING & DATABASES

0-7645-6852-3

0-7645-5784-X

Also available:
- A+ Certification For Dummies
  0-7645-4187-0
- Access 2003 All-in-One Desk Reference For Dummies
  0-7645-3988-4
- Beginning Programming For Dummies
  0-7645-4997-9
- C For Dummies
  0-7645-7068-4
- Firewalls For Dummies
  0-7645-4048-3
- Home Networking For Dummies
  0-7645-42796

- Network Security For Dummies
  0-7645-1679-5
- Networking For Dummies
  0-7645-1677-9
- TCP/IP For Dummies
  0-7645-1760-0
- VBA For Dummies
  0-7645-3989-2
- Wireless All In-One Desk Reference For Dummies
  0-7645-7496-5
- Wireless Home Networking For Dummies
  0-7645-3910-8